T0269614

Physical Sciences

From ancient times, humans have tried to understand the workings of the world around them. The roots of modern physical science go back to the very earliest mechanical devices such as levers and rollers, the mixing of paints and dyes, and the importance of the heavenly bodies in early religious observance and navigation. The physical sciences as we know them today began to emerge as independent academic subjects during the early modern period, in the work of Newton and other 'natural philosophers', and numerous sub-disciplines developed during the centuries that followed. This part of the Cambridge Library Collection is devoted to landmark publications in this area which will be of interest to historians of science concerned with individual scientists, particular discoveries, and advances in scientific method, or with the establishment and development of scientific institutions around the world.

The Arrangement of Atoms in Space

Awarded the first Nobel Prize in Chemistry in 1901 for his work on chemical dynamics and on osmotic pressure in solutions, the Dutch scientist Jacobus Henricus van 't Hoff (1852–1911) was also a pioneer in the field of stereochemistry – the three-dimensional analysis of chemical structures. This 1898 publication is based on the revised and expanded German translation of his *Dix années dans l'histoire d'une théorie* (1887), itself an updated version of his major work *La chimie dans l'espace* (1875). Translated and edited by the English chemist Arnold Eiloart, it covers the stereochemistry of carbon and nitrogen compounds, and contains an appendix on inorganic compounds by the Swiss chemist Alfred Werner (another future recipient of the Nobel Prize in Chemistry). Using experimental results, van 't Hoff shows how the varying spatial arrangement of similar compounds leads to differing chemical and optical behaviour.

Cambridge University Press has long been a pioneer in the reissuing of out-of-print titles from its own backlist, producing digital reprints of books that are still sought after by scholars and students but could not be reprinted economically using traditional technology. The Cambridge Library Collection extends this activity to a wider range of books which are still of importance to researchers and professionals, either for the source material they contain, or as landmarks in the history of their academic discipline.

Drawing from the world-renowned collections in the Cambridge University Library and other partner libraries, and guided by the advice of experts in each subject area, Cambridge University Press is using state-of-the-art scanning machines in its own Printing House to capture the content of each book selected for inclusion. The files are processed to give a consistently clear, crisp image, and the books finished to the high quality standard for which the Press is recognised around the world. The latest print-on-demand technology ensures that the books will remain available indefinitely, and that orders for single or multiple copies can quickly be supplied.

The Cambridge Library Collection brings back to life books of enduring scholarly value (including out-of-copyright works originally issued by other publishers) across a wide range of disciplines in the humanities and social sciences and in science and technology.

The Arrangement of Atoms in Space

JACOBUS HENRICUS VAN 'T HOFF
EDITED AND TRANSLATED BY
ARNOLD EILOART

CAMBRIDGE
UNIVERSITY PRESS

University Printing House, Cambridge, CB2 8BS, United Kingdom

Cambridge University Press is part of the University of Cambridge.
It furthers the University's mission by disseminating knowledge in the pursuit of education, learning and research at the highest international levels of excellence.

www.cambridge.org
Information on this title: www.cambridge.org/9781108080293

This edition first published 1898
This digitally printed version 2015

ISBN 978-1-108-08029-3 Paperback

ARRANGEMENT

OF

ATOMS IN SPACE

THE ARRANGEMENT

OF

ATOMS IN SPACE

BY

J. H. VAN 'T HOFF

SECOND REVISED AND ENLARGED EDITION

WITH A PREFACE BY JOHANNES WISLICENUS

Professor of Chemistry at the University of Leipzig

AND AN APPENDIX

STEREOCHEMISTRY AMONG INORGANIC SUBSTANCES

BY

ALFRED WERNER

Professor of Chemistry at the University of Zürich

TRANSLATED AND EDITED BY

ARNOLD EILOART

LONGMANS, GREEN, AND CO.
39 PATERNOSTER ROW, LONDON
NEW YORK AND BOMBAY
1898

PREFACE

TO

THE FIRST EDITION

THE first edition of this little book appeared in 1877, in the form of Dr. F. Herrmann's free rendering of my brochure, 'La Chimie dans l'Espace,' and Wislicenus, as long ago as that, helped the work by a warm recommendation.

As the original views still survive in Stereochemistry, this second edition presents once more a freely revised version of that brochure; but a section on nitrogen derivatives has been added. Besides this, in the part devoted to carbon, the greatly increased number of facts has been taken into account, and finally the amount of the rotation of active bodies has received special attention. Accordingly the book may serve as a reference book for stereochemistry and optical activity.

At the publishers' wish, I have studied brevity as far as compatible with thorough treatment.

J. H. VAN 'T HOFF.

AMSTERDAM: *February* 1894.

PREFACE

TO

THE SECOND EDITION

FOR this second edition of the 'Arrangement of Atoms in Space,' as for the first, the publishers and the author desire a short preface from my pen. This can have now no such purpose as in the case of Dr. Herrmann's edition. Then I had to address to German chemists a letter of recommendation in favour of the little-known hypothesis of a very young colleague; now the name of the author has a renown so high, based on such an extraordinary series of important and far-reaching researches, that my recommendation would be altogether superfluous for his book, even if the theory here set forth had not acquired for itself the position in chemistry which in fact it possesses.

Indeed, the old opposition to the principle has almost died out; where it still lives it is directed against the ultimate basis—against the Atomic Hypothesis itself—and does not deny that the doctrine of atomic arrangement in three dimensions is a logical and necessary stage, perhaps the final stage, in the chemical theory of atoms. For the most part the opposition is directed—often quite rightly—against special applications of the principle to the explanation of particular facts, leaving the principle itself untouched. That the hypothesis itself has proved its own justification—at least, as much as any other scientific theory—none can dispute.

It has already effected to the full all that can be effected by any theory; for it has brought into organic connection with the fundamental theories of chemistry facts which were before incomprehensible and apparently isolated, and has enabled us to explain them from these theories in the simplest way. By propounding to us new problems the hypothesis has stimulated empirical investigation on all sides; it has caused a vast accumulation of facts, has led to the discovery of new methods of observation, has become amenable to the test of experiment, and has at the same time started in our science a movement full of significance—in a certain sense, indeed, a new epoch.

How and to what extent the hypothesis has effected this, is told in this book, briefly, clearly, completely. The book is now not so much a new edition of the first German work, as a German revision of van 't Hoff's 'Dix Années dans l'Histoire d'une Théorie,' enriched by the growth of our knowledge during the last seven years. In this new form, also, the book will win many friends, and be a welcome guide to the comprehension of stereochemistry and to its already very extensive literature.

I may well be pardoned if I find an especial satisfaction in this new edition of van 't Hoff's pioneer publication. When it first appeared as 'La Chimie dans l'Espace' it bore as motto a sentence uttered by me as early as 1869.[1] I was then able to do something towards making known the new theory, and later to contribute to its development and to the experimental testing of it. Accordingly it is with great pleasure that I accept the honour of introducing the new revision, and send my thanks and regards to my honoured friend at Amsterdam.

JOHANNES WISLICENUS.

LEIPZIG: *April* 1894.

[1] *Ber.* **2**, 550, and especially p. 620.

CONTENTS

CHAPTER III

COMPOUNDS WITH SEVERAL ASYMMETRIC CARBON ATOMS

CHAPTER IV

CHAPTER V

THE UNSATURATED CARBON COMPOUNDS

CHAPTER VI

CHAPTER VII

NUMERICAL VALUE OF THE ROTATORY POWER

STEREOCHEMISTRY OF NITROGEN COMPOUNDS

APPENDIX

INTRODUCTION

EVERY time I write on stereochemistry a new name has to be added to complete the history of its development. In my 'Dix Années dans l'Histoire d'une Théorie' I mentioned Gaudin and his 'Architecture du Monde' (1873); then Meyerhoffer in his 'Stereochemie' added Paterno,[1] who in 1869 proposed to explain isomeric bromethylenes by a tetrahedral grouping round carbon; and Rosenstiehl,[2] who in the same year represented benzene by six tetrahedra; and now Eiloart, in his 'Guide to Stereochemistry,' goes back to Swedenborg's 'Prodromus Principiorum Rerum Naturalium sive Novorum Tentaminum Chymicam et Physicam Experimentalem geometrice explicandi.'[3] Certainly, then, we were not over-hasty, Le Bel and I, when we published our ideas (November and September 1874) in the 'Bulletin de la Société Chimique' and in the 'Voorstel tot Uitbreiding der Structuur-Formules in de Ruimte' respectively. That shortly before this

[1] *Giorn. di Scienze Naturali ed Econ.* vol. v., Palermo; *Gazz. Chim.* 1893, 35.

[2] *Bull. Soc. Chim.* **11**, 393. [3] Jan Oosterwyk, Amsterdam, 1721

we had been working together in Wurtz's laboratory was purely fortuitous; we never exchanged a word about the tetrahedron there, though perhaps both of us already cherished the idea in secret. To me it had occurred the year before, in Utrecht, after reading Wislicenus' paper on lactic acid. 'The facts compel us to explain the difference between isomeric molecules possessing the same structural formulæ by the different arrangement of their atoms in space': this was the sentence which remained in my memory, and which I have since used as a motto; on trying to refer to it I could not find it again, and so cannot give the reference here.

On the whole, Le Bel's paper and mine are in accord; still, the conceptions are not quite the same. Historically the difference lies in this, that Le Bel's starting point was the researches of Pasteur, mine those of Kekulé.

The researches of Pasteur had made plain the connection between optical activity and crystal-form, and had led to the idea that the isomers of opposite rotatory power correspond to an asymmetric grouping and to its mirrored image. Indeed, the possibility of a tetrahedral grouping was suggested.[1] Le Bel closely follows Pasteur, then, when he sees this grouping in the four atoms or radicals—inactive bodies all different—united to carbon.

My conception is, as Baeyer pointed out at the Kekulé festival, a continuation of Kekulé's law of the quadrivalence of carbon, with the added hypothesis

[1] *Leçons sur la Dissymétrie Moléculaire.*

that the four valences are directed towards the corners of a tetrahedron, at the centre of which is the carbon atom.

Practically our ideas, so far as they concern the asymmetric carbon, amount to the same thing—explanation of the two isomers by means of the tetrahedron and its image, disappearance of this isomerism when two groups become identical, through the resulting symmetry and identity of the two tetrahedra.

In the case of doubly linked carbon, however, there arises the possibilty of a difference. Here, too, four groups are connected, and Le Bel considers that *à priori* only so much is known about their position, that of the two pairs one pair lies nearer to one carbon, the other pair to the other carbon. It may happen, then, that ethylene derivatives may have no symmetry in their molecules—they may be active. Carrying out my tetrahedral grouping I concluded that the four groups are in one plane with the carbon, this being accordingly the plane of symmetry of all ethylene derivatives ; therefore no optical activity can occur. As regards this, Le Bel[1] at first altered his opinion in my favour, but later[2] altered it back again.

Of course, the facts must decide ; as, however, Liebermann informs me, specially for this edition, that bromocinnamic acid from active cinnamic acid dibromide is inactive, and Walden states that fumaric acid from active bromosuccinic acid is inactive, it

[1] *Bull. Soc. Chim.* 37, 300. [2] *Ibid.* [3], 7, 164, 1892.

appears that, in accordance with facts previously known, the simple conception of the tetrahedral grouping and of the development of structural chemistry to stereochemistry on these lines is still permissible.

STEREOCHEMISTRY OF CARBON

CHAPTER I

THE ASYMMETRIC CARBON ATOM

I. Statement of the Fundamental Conception

The molecule a stable system of material points.—When we arrive at a system of atomic mechanics the molecule will appear as a stable system of material points; that is the fundamental idea which continually becomes clearer and clearer when one is treating of stereochemistry; for what we are dealing with here is nothing else than the spatial—*i.e.* the real—positions of these points, the atoms.

I choose this fundamental conception as the starting point for this reason, that there is already evident in the rough outlines of this future system of atom mechanics a very considerable simplification, which I will here discuss.

One might suppose that the arrangement of the atoms in the molecule would be something like that in a system of planets, equilibrium being maintained by attraction and motion, by equality between centripetal and centrifugal force. I will try to show that we must exclude this motion; and this as a necessary consequence of simple thermodynamic considerations.

As the partial decomposition of salts containing water of crystallisation shows, and, generally, as the formula $\frac{d.l.K}{d.T} = \frac{q}{2T}$ requires, the alteration of any dissociation phenomenon with the temperature is always of this nature, that while on cooling the decomposition gradually becomes less, yet it ceases only at absolute zero ($T=0$). But this is as much as to say that the internal stability of the molecule is attained only at absolute zero, *i.e.* in the absence of all internal motion. Otherwise interaction with another molecule is an essential condition of equilibrium.

This law is seen to be perfectly general when we consider that every compound would undergo visible dissociation at a sufficiently high temperature, thus fulfilling the above conditions.

We may add, as a necessary consequence, that the state of things at absolute zero is to be explained solely by atomic mechanics, thermodynamics having nothing to do with this explanation, because thermodynamics comes into play only when, at a temperature above zero, dissociation begins; and we may add further that, to render equilibrium possible, instead of the centrifugal force which ordinarily acts, there must be a repulsion, for the action of matter (atoms) alone appears insufficient, and there must be something else, perhaps electricity.

For stereochemistry the above considerations are important as showing that motion of the atoms may for the present be neglected, the state of things being tacitly assumed to be as it would be at absolute zero. Indeed, the phenomena of isomerism are in a certain sense opposed to motion; they are certainly not a consequence thereof; for when the temperature rises they ultimately disappear, and become constantly more marked as it falls. He who chooses to assume motion, however, may conceive the motionless systems here to be described as the expression of the position of certain points about which the motion, doubtless a periodical motion, takes place.

Insufficiency of strüctural chemistry. The asymmetric carbon atom.—Everyone is now familiar with the fact, only occasionally observed in 1874, that the simple structural formulæ are insufficient to explain the existing cases of isomerism; and that, to consider first carbon-compounds of the type $C(R_1R_2R_3R_4)$—*i.e.* compounds in which four separate groups or atoms are combined with the carbon—an extra isomerism occurs when these four groups are different, and disappears if but two of them become the same. Assuming a fixed position of the groups round the carbon atom, only a tetrahedral grouping brings us to the same conclusion, as figs. 3 and 4 show: these become identical when R_3 and R_4 become the same; while this leaves the isomerism unaffected if we represent the formula in one plane (figs. 1 and 2).

To illustrate the matter with models we may use the cardboard tetrahedra, the different groups being represented by attaching caps made of coloured paper: *e.g.* R_1 white; R_2 yellow; R_3 black; R_4 red; to make the two tetrahedra alike an extra pair, say a pair of black caps, may be used, and may be placed on R_4, for instance. The Kekulé models, improved by v. Baeyer, and sold by Sendtner (Schillerstrasse 22, München), may be used for the same purpose.

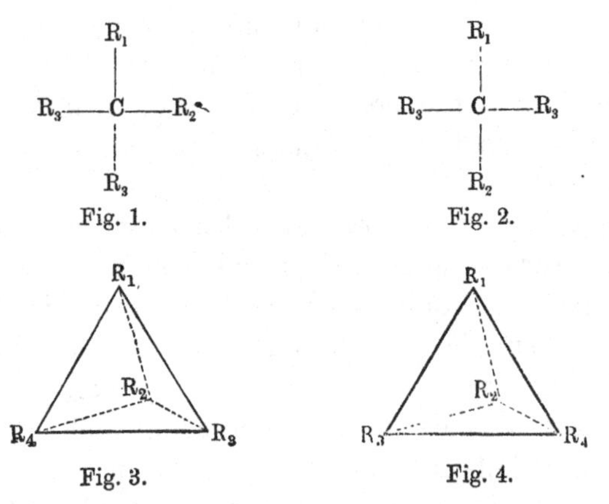

Fig. 1. Fig. 2. Fig. 3. Fig. 4.

A word as to the shape of the tetrahedra. If we wish to represent only the two possible formulæ given above, their peculiar lack of symmetry, their object-and-image relation, and the way they may be rendered identical, the regular tetrahedron with variously coloured corners quite suffices. But if the mechanics of the atoms is to be taken into account, we must admit, without making any hypothesis as to the nature of the forces acting, that in general these forces will be different between different groups, and the same between similar groups; and then the difference between fig. 3 and fig. 4 must be expressed in the dimensions also. The edges R_1 R_4, R_1 R_3, &c., will then be different in the two figures, but the corresponding dimensions in each, R_1 R_4 in fig. 3, R_1 R_4 in fig. 4, &c., will be equal. The two tetrahedra then express by their shape their object-and-image relation (so-called enantiomorphism), and at the same time a mechanical necessity is satisfied. It is now superfluous to vary the colours of the corners; but the way in which identity arises can now be shown only by two more models in which, in accordance with the fundamental requirement of mechanics, R_4 and R_3 take corresponding positions which are equally distant from the plane of symmetry now called into existence, and passing through R_1 R_2; for we now have

$$R_4\,R_1 = R_3\,R_1 \text{ and } R_4\,R_2 = R_3\,R_2.$$

II. Experimental Confirmation of the Fundamental Conception

A. CHARACTER OF THE ISOMERISM DUE TO THE ASYMMETRIC CARBON

Optical activity.—The isomerism expressed by the difference between the two enantiomorphous forms is characterised in the first place by this, that it is to be expected when the carbon is united to four different groups, and only then.

But in the second place all the molecular dimensions being equal in the two forms, we must expect a kind of isomerism distinguished by a near approach to identity. This state of things fully coincides with the facts, and may be summed up as follows.

All physical properties depending on molecular dimensions and attractions (mathematically speaking, on the quantities a and b of Van der Waals' theory) are identical in the two isomers; thus, sp. gr., crit. temp., maximal tension, boiling point, melting point, latent heat of fusion and vaporisation, &c. The same holds for the physical properties which manifest themselves as the expression of these fundamental quantities, on contact with other bodies, *e.g.* solubility.

As regards chemical properties we must expect exactly equal stability, the same speed of formation and of conversion in given reactions, equilibrium when equal quantities of each are present together, no heat of transformation when one is converted into the other, and accordingly equal heat of formation in both cases.

Finally, the only difference is due to the lack of symmetry, and this is manifested physically in the opposite optical activity, the so-called right- and left-handed rotation, shown by the isomers in the dissolved state—in the state, that is, when the rotation must arise from molecular, not from crystalline structure. It is important to note that a corresponding enantiomorphous structure causes the opposite activity in other cases also, as may be deduced empirically from active crystals, *e.g.* quartz, in which opposite rotatory power as regards light accompanies enantiomorphism of crystalline form. The same holds for elastic bodies wound in a right- or left-handed spiral, and finally for the active mica combinations of Reusch, formed of a pile of thin plates of binaxial mica placed at an angle of 60° one above another.[1] And Sarrau[2] has shown the theoretical necessity for this optical phenomenon in the case of asymmetric structures in general.

Crystalline form.—In the second place the asymmetry manifests itself crystallographically,[3] isomers due to asymmetric carbon showing an enantiomorphism of crystalline form corresponding to their molecular structure, as illustrated by the following woodcuts of right- and left-handed ammonium bimalate:—

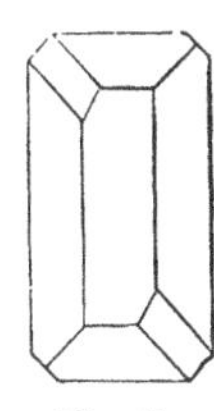

Fig. 5.

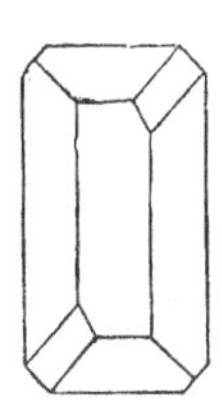

Fig. 6.

We may add that Soret[4] has shown this result to be a general necessity.

In the third place there is the difference in chemical and consequently in physiological properties.—The

[1] Wyrouboff, *Ann. de Chim. et de Phys.* [6], **8**, 340.
[2] *Journ. de Math. pures et appliquées* [2], **12**, 1867.
[3] But see Walden, *Ber.* **29**, 1692, and H. Traube's reply, *l.c.* 2446.
[4] *Arch. de Genève*, **24**, 592, 1890.

chemical identity emphasised above ceases directly the asymmetric isomers have to do with a substance which is itself asymmetric. And it is just in this case that mechanical reasons show the impossibility of equality in action (mathematically in the quantities *a* of Van der Waals for the two substances). The affinities are accordingly different, and doubtless the solubilities also; the resulting compounds have not the same composition, *e.g.* salts, as regards water of crystallisation; sometimes indeed one isomer can exist, while the other is incapable of existing. Finally the physiological action, particularly the nutritive value for the lower organisms, is altogether different for the two isomers, doubtless owing to the difference above mentioned, for in these organisms asymmetric bodies, *e.g.* proteids, play a great part. Also poisoning power, in the case of tartaric acid, and taste, in the case of asparagine, are different.

We may add that Pasteur[1] long ago expressed the view that the above-mentioned optical, crystallographical, chemical, and physiological properties must arise from an asymmetric grouping in the molecule; indeed, he even mentioned the tetrahedron as a possibility.

'Are the atoms of tartaric acid arranged along the spiral of a right-handed screw, or are they situated at the corners of an irregular tetrahedron, or have they some other asymmetric grouping? Time must answer the question. But of this there is no

[1] *Leçons de Chimie*, 1860, 25.

doubt, that the atoms possess an asymmetric arrangement like that of an object and its mirrored image. Equally certain is it that the atoms of the left-handed acid possess just the opposite asymmetric arrangement.'

B. OBSERVED COINCIDENCE OF OPTICAL ISOMERISM WITH THE PRESENCE OF ASYMMETRIC CARBON

Enumeration of the active compounds.—In order now to show that the above-mentioned properties really accompany the asymmetric carbon atom wherever it occurs, we may confine our attention to the optical activity, since all the above-mentioned peculiarities, crystallographical, chemical, and physiological, regularly coincide therewith. Another simplification: it is no matter whether both isomers have been found or not, since if a compound has been found which in solution rotates to the right (*e.g.*), it is perfectly certain that the corresponding compound of opposite activity will sooner or later be discovered.

We will therefore simply enumerate the active bodies, of known structure, indicating the asymmetric carbon by italics. In order to give these important data as completely as possible without occupying too much space I will here confine myself to the active compounds containing only one asymmetric carbon, as the others will be discussed later; and in each group only the principal members (and not, *e.g.*, salts and esters) will be included, since these derivatives also will bc treated of specially.

1. **Compounds with three carbon atoms.**—(a_D indicates the directly observed rotation for sodium light; $[a]_D$ is the so-called specific rotation, *i.e.* calculated for one decimetre and unit density.)

Propylene glycol,[1] $CH.OH.CH_3.CH_2OH$ $a_D = -5°$ for 22 cm.

Propylene oxide,[1] $CH.CH_3.OCH_2$ $a_D = +1°$ for 22 cm.

Propylene diamine, $CH.CH_3.NH_2.CH_2NH_2$.[2]

Ethylidene lactic acid,[3] $CH.OH.CH_3.CO_2H$. Rotation in aqueous solution varying much with time and concentration (c).

Maximal value $[a]_D = +3°$ ($c = 7{\cdot}38$). The somer of opposite activity has also been obtained.[4]

Lactid,[3] $CH.CH_3.CO.O$ $[a]_D = -86°$.

Cystine,[5] $C.SH.NH_2.CH_3.CO_2H$ $[a]_D = -8°$.

Glyceric acid,[6] $CH.OH.CH_2OH.CO_2H$. Rotation in aqueous solution varying much with time and concentration. Both isomers have been obtained.

2. **Compounds with four carbon atoms.**

Butyl alcohol,[7] $CH.OH.CH_3.C_2H_5$.

[1] Le Bel, *Bull. Soc. Chim.* [2], **34**, 129.

[2] Baumann, *Ber.* **28**, 1177.

[3] Wislicenus, *Ann.* **167**, 302.

[4] Nencki and Sieber, *Ber.* **22**, Ref. 695; Schurdinger, *Chem. Soc. J. Abstr.* 1891, p. 666; Purdie and Walker, *Trans.* 1892, p. 754; Lewkowitsch, *Ber.* **16**, 2720; Linossier, **24**, Ref. 660.

[5] Baumann, *Zeitschr. f. Physiol. Chem.* **8**, 305.

[6] Lewkowitsch, *l.c.*; Frankland and Appleyard, *Chem. Soc. J. Trans.* 1893, p. 296.

[7] Combes and Le Bel, *Chem. Soc. J. Abstr.* 1893, p. 246.

Oxybutyric acid,[1] $CH.OH.CH_3.CH_2CO_2H$ $[a]_D = -21°$.

Malic acid, $CH.OH.CO_2H.CH_2CO_2H$. Rotation in aqueous solution varying much with the concentration. $[a]_D = -2·3°$ ($c = 8·4$); $[a]_D = +3·34°$ ($c = 70$).[2]

The isomer of opposite activity has also been obtained.[3]

Chlorosuccinic acid, $CHCl.CO_2H.CH_2CO_2H$[4] $[a]_D^{21} = +20°$ ($c = 3·2$ to 16).

Methoxysuccinic acid,

$$CH.OCH_3.CO_2H.CH_2CO_2H$$ [5]

$[a]_D^{18} = 33°$ ($c = 5·5$ to $10·8$).

Both isomers were obtained.

Ethoxysuccinic acid, $CH.OC_2H_5.CO_2H.CH_2CO_2H$[6] $[a]_D^{18} = 33°$ ($c = 5·6$ to $11·2$).

Propoxysuccinic acid,

$$CH.OC_3H_7.CO_2H.CH_2CO_2H \ ^{7} \ [a]_D^{12} = 36·2°.$$

Isopropoxysuccinic acid,

$$CH.OCH(CH_3)_2.CO_2H.CH_2CO_2H.$$ [8]

Aspartic acid, $CH.NH_2.CO_2H.CH_2CO_2H$ $[a]_D = -4°$ to $-5°$[9] in aqueous solution. The isomer of opposite activity is also known.[10]

[1] Minkowski and Külz, *Ber.* **17**, Ref. 334, 534, 535.

[2] Schneider, *Ann.* **207**, 257.

[3] Bremer, *Bull. Soc. Chim.* **25**, 6; Piutti, *Ber.* **19**, 1693.

[4] Walden, *Ber.* **26**, 215.

[5] Purdie and Marshall, *Chem. Soc. J. Trans.* 1893, p. 217.

[6] Purdie and Walker, *l.c.* p. 229.

[7] Purdie and Bolam, *ibid.* 1895, p. 955; Cook, *Ber.* **30**, 294.

[8] Purdie and Lander, *Chem. Soc. Proc.* 1896, p. 221.

[9] Becker, *Ber.* **14**, 1031.

[10] Piutti, *Compt. Rend.* **103**, 134; Marshall, *Chem. Soc. Trans.* 1896, p. 1023.

Malamide, $CH.OH.CONH_2.CH_2CONH_2$.

Asparagine, $CH.NH_2.CO_2H.CH_2CONH_2$ $[a]_D = -8° \text{ to } -5°$[1] in aqueous solution. The isomer of opposite activity is also known.[2]

Uramidosuccinamide,

$$CH.NHCONH_2.CO_2H.CH_2CONH_2,$$

also obtained in the two modifications.[2]

3. **Compounds with five carbon atoms.**

Amyl alcohol, $CH.CH_3.C_2H_5.CH_2OH$ $[a]_D^{22} = -5°$.[3] The isomer of opposite activity is also known.[4] All the derivatives are mentioned later.

Valeric acid, $CH.CH_3.C_2H_5.CO_2H$ $[a]_D = +17° 30'$.[5]

Amyl alcohol, $CH.OH.CH_3.C_3H_7$ $a_D = -8° 7'$ for 22 cm.[6]

Amyl iodide, $CHI.CH_3.C_3H_7$ $a_D = +1° 8'$ for 22 cm.[6]

Amyl chloride, $CHCl.CH_3.C_3H_7$ $a_D = -0° 5'$ for 20 cm.[7]

Oxyglutaric acid, $CH.OH.CO_2H.C_2H_4CO_2H$ $[a]_D = -2°$[8] in aqueous solution.

Methylmalic acid, $C.OH.CH_3.CO_2H.CH_2CO_2H$.[9]

[1] Becker, *Ber.* **14**, 1031.

[2] Piutti, *Compt. Rend.* **103**, 134; Marshall, *Chem. Soc. Trans.* 1896, p. 1023.

[3] Rogers, *Chem. Soc. J. Trans.* 1893, p. 1130.

[4] Le Bel, *Compt. Rend.* **87**, 213; *cf.* Schütz and Marckwald, *Ber.* **29**, 52.

[5] Taverne, *Rec. des Trav. Chim. des Pays-Bas*, 1894, p. 201; Schütz and Marckwald, *l. c.*

[6] Le Bel, *Bull. Soc. Chim.* **33**, 106.

[7] Guye, *Thèses*, 1891, p. 55.

[8] Ritthausen, *J. f. prakt. Chem.* [2], **5**, 354; Scheibler, *Ber.* **17**, 1728.

[9] Le Bel, *Bull. Soc. Chim.* [3], **11**, 292.

Glutamic acid, $CHNH_2.CO_2H.C_2H_4CO_2H$ $[\alpha]_D = +35°$ [1] in dilute nitric acid.

4. **Fatty substances with more than five carbon atoms.** Hexyl alcohol,[2] $CH.CH_3.C_2H_5.CH_2CH_2OH$ $[\alpha]_D = 8°$.

Hexylic acid,[2] $CH.CH_3.C_2H_5.CH_2CO_2H$ $[\alpha]_D = 9°$. The isomer of opposite activity has been prepared from amyl alcohol.[3]

Hexyl alcohol[4] $\begin{cases} CH.OH.CH_3.C_4H_9, \text{ left-handed.} \\ CH.OH.C_2H_5.C_3H_7, \text{ right-handed.} \end{cases}$

Hexyl chloride,[4] $CH.Cl.C_2H_5.C_3H_7$, left-handed.

Hexyl iodide,[4] $CH.I.C_2H_5.C_3H_7$, right-handed.

Leucine,[5] $CH.CH_2CH(CH_3)_2.NH_2.CO_2H$ $[\alpha]_D = +18°$ in hydrochloric acid solution.[6] The isomer of opposite activity has also been discovered.[7]

Ethyl amyl,[8] $CH.CH_3.C_2H_5.C_3H_7$ $\alpha_D = +5°$ for 20 cm.

5. **Pyridine derivatives.**

α-Pipecoline = α-methylpiperidine[9] $[\alpha]_D = 35°$.

α-Ethylpiperidine[9] $[\alpha]_D = 7°$.

Conine = α-propylpiperidine[9] $[\alpha]_D = 14°$.

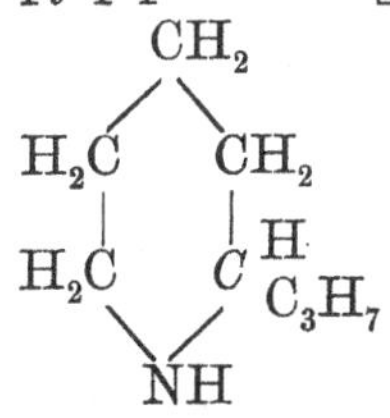

[1] Ritthausen, *J. f. prakt. Chem.* [1], **107**, 238.
[2] Van Romburgh, *Rec. des Trav. Chim. des Pays-Bas*, **6**, 150.
[3] Wurtz, *Ann. Chim. Phys.* [3], **51**, 358.
[4] Combes and Le Bel, *Chem. Soc. J. Abstr.* 1893, p. 246.
[5] Schulze and Likiernik, *Ber.* **24**, 669 ; **26**, Ref. 500.
[6] Mauthner, *Zeitschr. f. physiol. Chem.* **7**, 222.
[7] Schulze, Barbieri, and Bosshard, *ibid.* **9**, 103.
[8] Just, *Ann.* **220**, 157.
[9] Ladenburg, *Ber.* **19**, 2584, 2975.

Copellidine = methylethylpiperidine.[1]
Methylconine[2] $[a]_D = 81 \cdot 33°$.
Nicotine[3] $[a]_D = -161°$.

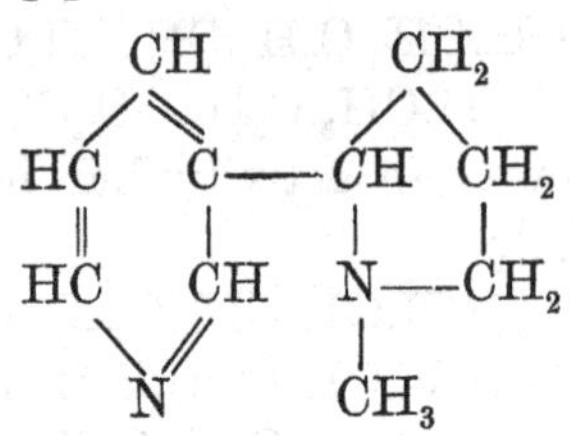

6. **Aromatic compounds.**

Mandelic acid,[4] *C*H.C_6H_5.OH.CO_2H $[a]_D = \pm 156°$.

Tropic acid,[5] *C*H.C_6H_5.CH_2OH.CO_2H $[a]_D = 71°$ in aqueous solution. Both isomers were obtained.

Phenylcystine,[6] *C*.CH_3.NH_2.SC_6H_5.CO_2H $[a]_D = -4°$.

Bromophenylcystine,[6] *C*.CH_3.SC_6H_4Br.NH_2.CO_2H.

Phenylbromomercapturic acid[6] $[a]_D = -7°$.
C.CH_3.SC_6H_4Br.NH($COCH_3$).CO_2H.

Phenylamidopropionic acid,[7]
*C*H.NH_2.$CH_2C_6H_5$.CO_2H.

Tyrosine,[8] CH.NH_2.CH_2.C_6H_4OH.CO_2H $[a]_D = -8°$.

Isopropylphenylglycollic acid,[9] *C*H.OH.$C_6H_4C_3H_7$.CO_2H. Both isomers were prepared. $[a]_D = 135°$.

[1] Levy and Wolffenstein, *Ber.* **28**, 2270; **29**, 43, 1959.
[2] Wolffenstein, *ibid.* **27**, 2614.
[3] Ladenburg, *ibid.* **26**, 293.
[4] Lewkowitsch, *ibid.* **16**, 1565, 2721.
[5] Ladenburg, *ibid.* **22**, 2590.
[6] Baumann, *ibid.* **15**, 1401, 1731.
[7] Schulze and Nägeli, *Zeitschr. f. physiol. Chem.* **11**, 201.
[8] Mauthner, *Wien. Akad. Ber.* [2], **85**, 882.
[9] *Ber.* **26**, Ref. 89.

Leucinephthaloylic acid,[1]
$CH.C_4H_9.NHCOC_6H_4(CO_2H).CO_2H$.
Phthalylamidocaproïc acid,[1]
$CH.C_4H_9.N(C_2O_2)C_6H_4.CO_2H$.

Limonene.[2]	Carvol.[3]	Camphor.[4]
$[a]_D = \pm 105°$	$[a]_D = \pm 62°$	$[a]_D = \pm 55°$
C_3H_7	C_3H_7	C_3H_7
C	C	C
HC CH	HC CH	H_2C CH
HC CH_2	HC CO	H_2C CO
C	*C*	*C*
HCH_3	HCH_3	HCH_3

Tetrahydronaphthylenediamine[5] $[a]_D = -7°$ and $+8°$.

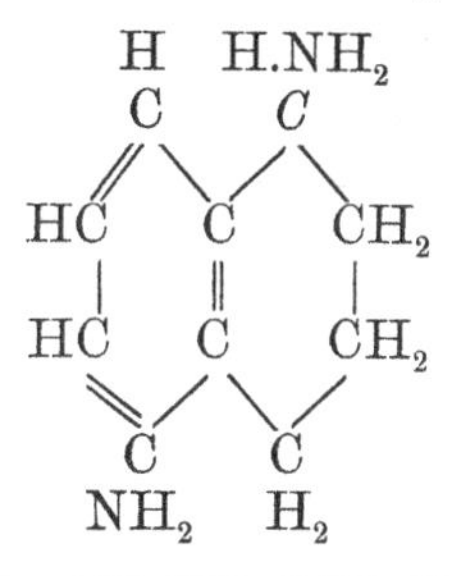

Phenyl amyl,[6] $CH.CH_3.C_2H_5.CH_2C_6H_5$ $a_D = 1° 4'$.

Predictions of activity confirmed.—Thus in all active bodies the asymmetric carbon occurs; indeed,

[1] Reese, *Ann.* **242**, 9; *Ber.* **21**, 277.
[2] *Ber.* **21**, 166.
[3] *Ibid.* **20**, 486, 2071.
[4] Landolt, *Opt. Drehungsvermögen*, p. 83.
[5] Bamberger, *Ber.* **23**, 291.
[6] Guye, *Thèses*, 1891.

in many cases activity was first suspected from the constitution, and subsequently discovered. This was the case with leucine, tyrosine, cystine, propyleneglycol, glyceric and mandelic acids, secondary butyl-, amyl-, and hexyl-alcohol, isopropylphenylglycollic acid, hydronaphthalenediamine, &c.

Doubtful statements.—A fact which inspires special confidence is that in seven cases the supposed activity of bodies containing no asymmetric carbon atom has been disproved.

Propyl alcohol, $CH_3CH_2CH_2OH$.[1]

Styrolene, $C_6H_5CHCH_2$.[2]

Trimethylethylstibineiodide, $(CH_3)_3C_2H_5SbI$.[3]

β-Picoline = β-methylpyridine.[4]

Papaverine.[5]

Chlorofumaric and chloromaleïc acids,[6]

$$CO_2HCClCHCO_2H.$$

It must then be considered doubtful whether oxypyroracemic acid[7] with the constitution ascribed to it, $CO_2H.COCH_2OH$, really possesses the activity discovered by Will.

[1] Chancell, *Compt. Rend.* **68**, 659, 726. Inactive according to a private communication from Henninger.

[2] Berthelot, *ibid.* **63**, 518; van 't Hoff, *Maandblad voor Natuurwetenschappen*, **6**, 72; *Ber.* **9**, 5; Krakau, *Ber.* **11**, 1259; Weger, *Ann.* **221**, 68.

[3] Friedländer, *Journ. f. prakt. Chem.* **70**, 449; Le Bel, *Bull. Soc. Chim.* **27**, 444.

[4] Hesekiel, *Ber.* **18**, 3091; Landolt, *ibid.* **19**, 157.

[5] Hesse, *Ann.* **176**, 198; Goldschmidt, *Wien. Akad. Ber.* January 1888.

[6] Perkin, *Chem. Soc. Journ.* 1888, 695; van 't Hoff, *Ber.* **10**, 1620; Walden, *l.c.* **26**, 210, 508.

[7] *Ber.* **24**, 400

Disappearance of the activity in derivatives.—It is of special importance to note the activity in different groups of derivatives, for it is found that the activity regularly vanishes with *C*, the asymmetric carbon atom. This proof is specially pertinent, because Colson [1] has recently given prominence again to the conception of an active type or radical as the cause of rotation; this conception, however, lacks sufficient precision, the precision which renders it possible to decide beforehand in which cases this type vanishes.

In the amyl series, in the derivatives of active amyl alcohol, $H_3C(C_2H_5)\mathit{C}HCH_2OH$, the activity persists in the ethers and amyl sulphates, in the chloride, bromide, and iodide, in amylamine and its salts, in the aldehyde and in valeric acid, in diamyl; in short, in more than sixty compounds recently examined by Guye [2] and others. Unaided by the theory, one would be inclined to maintain that the activity exists in all the derivatives; but, relying on the theory, Le Bel [3] and Just [4] examined the nearest derivatives in which the asymmetric carbon is lacking, the former testing methylamyl, $(H_5C_2)_2CHCH_3$, and amylene, $H_3C(C_2H_5)CCH_2$, the latter amylhydride. No rotatory power could be detected in any of these three compounds.

In the derivatives of tartaric acid the same

[1] *Étude sur la Stéréochimie*, 1892; *Journ. de Pharm. et de Chimie*, 1893.

[2] *Thèses*, Paris, 1891; Walden, *Zeitschr. physik. Chem.* **15**, 638; I. Welt, *Compt. Rend.* **119**, 885; Guye and Chavanne, *l.c.* **119**, 906; **120**, 452.

[3] *Bull. Soc. Chim.* [2], **25**, 565.

[4] *Ann.* **220**, 146.

peculiarity occurs. Starting with the right-handed acid we find the rotatory power preserved in the salts and esters, in tartramic acid and tartramide; in short, in forty-one derivatives recently enumerated by Guye, in malic acid, its salts, its esters, and its amide. But Pasteur himself did not suspect that the activity would disappear in succinic acid,[1] $CO_2HCH_2CH_2CO_2H$, obtained by the reduction of malic acid; the same holds for chlorofumaric acid,[2] $CO_2HCClCHCO_2H$, obtained by treating tartaric acid with phosphorus pentachloride.

Starting from malic acid in the contrary direction we have active methoxy- and ethoxy-succinic acids, chlorosuccinic acid, asparagine, aspartic acid with the two series of salts, uramidosuccinic acid;[3] but the succinic acid made from asparagine is inactive.

Further confirmation is afforded by the following isolated cases, which will find an application later.

Oxalic acid made from active sugar[4] or active tartaric acid[5] is inactive; so is furfurol from active arabinose or xylose.[5]

Active phenylcystine gives on treatment with baryta inactive phenylmercaptan.[6]

The active oxybutyric acid of Minkowski and Külz gives an inactive crotonic acid.[7]

[1] Bremer and van 't Hoff, *Ber.* **9**, 215.
[2] Van 't Hoff, *ibid.* **10**, 620; Walden, *l.c.* **26**, 210.
[3] Piutti, *Compt. Rend.* **103**, 134.
[4] Van 't Hoff, *Ber.* **10**, 620. [5] *Ibid.* 1620.
[6] Baumann and Preusse, *l.c.*
[7] Deichmüller, Szymansky, and Tollens, *Ann.* **228**, 95.

Among observations of this kind those cases in which compounds without the asymmetric atom are obtained by fermentation—*i.e.* by the action of living organisms—deserve special attention, because this action specially favours the formation of active compounds. When therefore in these circumstances an inactive body is formed from an active one, it is surely very probable that its inactivity arises from its constitution being incompatible with rotatory power. For this reason we mentioned in 1875 the inactivity of ethyl-, propyl-, butyl-, and amyl-alcohols, which result from the fermentation of active carbohydrates.

Succinic acid [1] made by fermentation of active malate and tartrate of calcium, of asparagine, and of starch, is inactive. Further, Beyerinck, to whom I am indebted for these samples of succinic acid (made by Fritz), has placed at my disposal some ethylacetate prepared by fermenting active maltose. Van Deventer showed that this was inactive.

Finally, one might add all inactive vegetable products, which for the most part are made from active material under the influence of the organism. The inactivity of citric acid, *e.g.*, rendered probable the formula $CO_2H.CH_2COH(CO_2H)CH_2CO_2H$. This was pointed out long ago and has since been proved.

Does any difference in the groups attached to the carbon suffice to cause activity?—This question arose in the first edition. Some reserve was still necessary as long as cases were lacking in which even the dif-

[1] *Ber.* **10**, 1620; **11**, 142; **12**, 474.

ference between halogen and hydrogen—*i.e.* between the simplest possible groups of only one atom—sufficed for activity.[1] There was room for doubt, because in many cases the activity was lost on substituting one group for another—*e.g.* chlorine for hydroxyl—although all the groups were still different. Thus the following inactive halogen derivatives were obtained from active bodies :—

Bromosuccinic acid from malic acid.[2]

Dichlorosuccinic acid from tartaric acid.[3]

Iodohexyl from mannite.[4]

Phenyl-brom- and chlor-acetic acids from phenylglycollic acid.[5]

Isopropylphenylchloracetic acid from isopropylphenylglycollic acid.[6]

Since then, however, the following thoroughly decisive cases have become known in which the activity is retained or present.

Chloro- and bromo-propionic acids from lactic acid.[7]

Chlorosuccinic acid from malic acid.[3]

Chloro- and bromo-malic acids from tartaric acid.[7]

Iodide and chloride of secondary amyl alcohol, $\mathit{C}H.OH.CH_3.C_3H_7$.[8]

Iodide and chloride of secondary hexyl alcohol, $\mathit{C}H.OH.C_2H_5.C_3H_7$.[9]

[1] See Guye's work below.
[2] *Ann.* **130**, 172 ; *Ber.* **24**, 2687.
[3] Walden, *Ber.* **26**, 212.
[4] *Ann.* **135**, 130.
[5] *J. Chem. Soc.* **59**, 71 ; *Proc.* 1891, 152.
[6] Fileti, *Gazz. Chim.* [2], **22**, 395.
[7] Walden, *Ber.* **28**, 1287.
[8] Le Bel, *Bull. Soc. Chim.* [2], **33**, 106 ; Guye, *Thèses*, 1891.
[9] Combes and Le Bel, *J. Chem. Soc. Abstr.* 1893, 246.

Hexachlorhydrin of mannite,

$CH_2Cl(CHCl)_4CH_2Cl$.[1]

Cinnamic acid dibromide and dichloride,[2]

$C_6H_5(CHBr)_2CO_2H$.

Finally, the above-mentioned phenyl-brom- and chlor-acetic acids, which had been known only in the inactive form, were obtained active from mandelic acid.

The unexpected occurrence of the inactive derivatives will be explained presently.

TRANSLATOR'S NOTE

A final restriction has yet to be acknowledged. At present we do not know a single active molecule containing less than two carbon atoms united with the asymmetric carbon.

Thus no activity has been observed in the following compounds :—

Containing one carbon atom.

Chlorobromomethanesulphonic acid, $CHClBr.SO_3H$.

Containing two carbon atoms.

Bromnitroethane, $CHBr.NO_2.CH_3$.
Sodiumnitroethane, $CHNa.NO_2.CH_3$.
Aldehyde ammonia, $CH.OH.NH_2.CH_3$.
Chloralsulphydrate, $CH.OH.SH.CCl_3$.
Chloral alcoholate and hydrocyanide.
Bromoglycollic acid, $CH.OH.Br.COOH$.
Hydrogen silver fulminate, $CHAg.NO_2.CN$.
Ethylidene iodobromide, $CHIBr.CH_3$.

[1] Mourgues, *Compt. Rend.* **111**, 112.
[2] Liebermann, *Ber.* **26**, 245, 833.

Ethylidenemethethylate, CH.OCH$_3$.OC$_2$H$_5$.CH$_3$.
Ethylidenechlorosulphinic acid, CHCl.SO$_2$H.CH$_3$.
Chlorethylidene oxide, CH$_3$.CHCl.O.ClHC.CH$_3$.

The verdict of observation, then, up to the present time, is that an asymmetric carbon alone is not sufficient to cause optical activity, but that the presence of the group C.C.C is essential.[1] It seems probable, however, that the inactivity of the molecules just mentioned is due to an intramolecular transformation, favoured by the mobility of the small radicals attached to the asymmetric carbon. The same thing is observed in the case of asymmetric nitrogen (p. 181 *post*).

[1] Compare Möller, *Cod Liver Oil and Chemistry*, p. 462 (London : Möller).

CHAPTER II

DIVISION OF THE INACTIVE MIXTURE[1]

INACTIVITY OF COMPOUNDS CONTAINING AN ASYMMETRIC CARBON ATOM

EVERY active compound, then, occurring in the two characteristic isomers contains an asymmetric carbon atom; on the other hand, there are many substances which possess this peculiar constitution and yet show no activity; indeed, they are perfectly certain to be inactive when prepared in the laboratory from inactive substances.

From the first, Le Bel and I considered this difficulty to be merely apparent. The exactly corresponding internal constitution of the two isomers, $CR_1R_2R_3R_4$, demands that when they are formed from $CR_1R_2R_3R_3$ (where the similar groups R_3 occupy exactly corresponding positions on each side of the plane of symmetry passing through CR_1R_2), the reaction should proceed with equal velocity in two directions; the product will consist of equal quantities of two isomers, one resulting from the conversion of the one R_3 group into R_4, the other

[1] Compare Chr. Winther, *Ber.* **28**, 3000. Tables showing the results so far obtained and a theoretical explanation are given.

from the conversion of the other R_3 group into R_4. Thus we get an inactive mixture, and a mixture which, owing to the complete agreement in the chemical and physical properties of the components, can be separated only by special means. If we add that, on the other hand, the two isomers, like right- and left-handed tartaric acid, may join together to form a so-called racemic compound, everything justifies the expectation that the isomers may be obtained from the product of this reaction, as Pasteur obtained tartaric from racemic acid.

And this has gradually been done in more than thirty cases. We have now, then, to describe the methods, which may be briefly indicated as—

Division by the use of active compounds; division by the use of organisms; spontaneous division; proof of divisibility by synthesis of the inactive mixture or compound.

1. DIVISION BY THE USE OF ACTIVE COMPOUNDS

This method was based on the observation of Pasteur that when a solution of racemic acid is neutralised with (active) cinchonine, the salt of the left-handed tartaric acid is the first to crystallise out. Since that time the method has been employed with success in many cases, and seems advantageous for procuring a large quantity of pure substance; but it is limited to the division of acids and bases, because other active bodies lack the requisite combining power.

Substance	Means	Author
Racemic acid, $CO_2H\mathit{C}HOH\mathit{C}HOHCO_2H$	Cinchonine	Pasteur.
Malic acid, $CO_2H\mathit{C}HOHCH_2CO_2H$	Cinchonine	Bremer, 'Ber.' **13**, 351; 'Rec. des Trav. Chim. des Pays-Bas,' **4**, 180.
α-Oxybutyric acid, $CO_2H\mathit{C}H.OH.C_2H_5$	Brucine	Guye and Chavanne, 'Compt. Rend.' **120**, 565, 632.
Pyrotartaric acid, $CO_2H\mathit{C}H.CH_3.CH_2CO_2H$	Strychnine	Ladenburg, 'Ber.' **28**, 1171.
Mandelic acid, $C_6H_5.\mathit{C}HOHCO_2H$	Cinchonine	Lewkowitsch, 'Ber.' **16**, 1573.
i-Mannonic acid, $CH_2OH(\mathit{C}HOH)_4CO_2H$	Strychnine, Morphine	Fischer, 'Ber.' **23**, 379.
i-Galactonic acid, $CH_2OH(\mathit{C}HOH)_4CO_2H$	Strychnine	Fischer, 'Ber.' **25**, 124.
Phenylbromolactic acid, $C_6H_5\mathit{C}HBr\mathit{C}HOHCO_2H$	Cinchonine	Purdie and Marshall, 'Chem. Soc. J. Trans.' 1893, 218.
Ethoxysuccinic acid, $CO_2H\mathit{C}H(OC_2H_5)CH_2CO_2H$	Cinchonine	Purdie and Walker, 'Chem. Soc. J. Trans.' 1892, 754.
Lactic acid, $CH_3\mathit{C}HOHCO_2H$	Strychnine	Purdie and Walker, 'Chem. Soc. J. Trans.' 1892, 754.
Isopropylphenylglycollic acid, $C_3H_7C_6H_4\mathit{C}HOHCO_2H$	Quinine, Cinchonine, Codeine	Fileti, 'Gazz. Chimica,' [2], **22**, 395.
Cinnamic acid dibromide $C_6H_5(\mathit{C}HBr)_2CO_2H$	Strychnine	Erlenmeyer, jun., and Lothar Meyer, jun., 'Ann.' **271**, 137; 'Ber.' **25**, 3121.
Cinnamic acid dichloride	Strychnine	Liebermann, 'Ber.' **26**, 833.
Phenyldibromobutyric acid	Brucine	L. Meyer, jun., and O. Stein, 'Ber.' **27**, 890.

Substance	Means	Author
Nitrosohexahydroquinolic acid	Strychnine	Besthorn, 'Ber.' **28**, 3156.
Conine, $HNC_5H_9C_3H_7$	Tartaric acid	Ladenburg, 'Ber.' **19**, 2975.
α-Pipecoline, $HNC_5H_9CH_3$	Tartaric acid	Ladenburg, 'Ber.' **19**, 2975.
α-Ethylpiperidine, $HNC_5H_9C_2H_5$	Tartaric acid	Ladenburg, 'Ber.' **19**, 2975.
1, 5-Tetrahydronaphthylene-diamine	Tartaric acid	Bamberger, 'Ber.' **23**, 291.
Diphenyldiethylenediamine	Tartaric acid	Feist and Arnstein, 'Ber.' **28**, 3169.
Copellidine (hydroaldehyde-collidine)	Tartaric acid	Levy and Wolffenstein, 'Ber.' **28**, 2270; **29**, 43.

2. DIVISION BY THE USE OF ORGANISMS

In this case it may be said that the division is due to the same causes as in the last—namely, to the different deportment of the active isomers towards the active compounds (proteids) of the organism. The origination of this method is also due to Pasteur, who observed that a dilute solution of ammonium racemate with a trace of phosphate leaves, after the growth of penicillium, finally a solution of the left-handed salt.

For stereochemical purposes the method has the advantage that it is not limited to acids and bases; on the other hand, one of the isomers is lost, whereas the former method yielded both. Moreover the preparation of a pure product is not so easy, because continued vegetation often destroys the other isomer also. The process seems specially suitable in cases where a qualitative test of possible activity is required; accordingly it was used in all Le Bel's investigations. In this way we may divide—

Substance	Means and Product	Author
Racemic acid, $CO_2HCHOHCHOHCO_2H$	Penicillium, L-tartaric acid	Pasteur.
Amyl alcohol, $CH.CH_3.C_2H_5CH_2OH$	Penicillium, R-alcohol	Le Bel, 'Compt. Rend.' **87**, 213.
Amyl alcohol, $CHOHCH_3C_3H_7$	Penicillium, L-alcohol	Le Bel, 'Compt. Rend.' **89**, 312.
Butyl alcohol, $CHOHCH_3C_2H_5$	Penicillium, L-alcohol	Combes and Le Bel, 'Bull. Soc. Chim.' [3], **7**, 551.
Hexyl alcohol, $CHOHCH_3C_4H_9$	Penicillium, L-alcohol	Combes and Le Bel, 'Bull. Soc. Chim.' [3], **7**, 551.
Butyleneglycol, $CHOHCH_3CH_2OH$	Bacterium termo, L-alcohol	Le Bel, 'Compt. Rend.' **92**, 532.
Mandelic acid, $CH.OH.C_6H_5.CO_2H$	Aspergillus, Mucor, Penicillium, R-acid;	Lewkowitsch, 'Ber.' **15**, 1505.
	Saccharomyces ellipsoïdeus, Schizomycetes, L-acid	Lewkowitsch, 'Ber.' **16**, 1569.
Glyceric acid, $CHOH.CH_2OH.CO_2H$	Penicillium, L-acid; Bacillus ethaceticus, R-acid	Lewkowitsch, 'Ber. **16**, 2721.
Lactic acid, $CH.OH.CH_3.CO_2H$	Penicillium, L-acid [?]	Lewkowitsch, 'Ber. **16**, 2721.
	R-acid	Linossier, 'Bull. Soc. Chim.' [3], **6**, 10; Schardinger, 'Mon. für Chem.' **11**, 545.

Substance	Means and Product	Author
Leucine, *C*H.NH$_2$.C$_4$H$_9$.CO$_2$H	Penicillium, left-handed in hydrochloric acid	Schulze and Bosshard, 'Ber.' **18**, 388.
Glutamic acid, *C*H.NH$_2$.CO$_2$H.C$_2$H$_4$CO$_2$H	Penicillium, left-handed in hydrochloric acid	Schulze and Bosshard, 'Ber.' **18**, 388.
Aspartic acid, *C*H.NH$_2$.CO$_2$H.CH$_2$CO$_2$H	Penicillium, left-handed in hydrochloric acid	Schulze and Bosshard, *l.c.*; Engel, 'Compt. Rend.' **106**, 1734.
α-Acrose, CH$_2$OH(*C*HOH)$_3$COCH$_2$OH	Beer yeast, L-fructose	Fischer, 'Ber.' **23**, 389.
i-Mannose, CH$_2$OH(*C*HOH)$_4$COH	Beer yeast, L-mannose	Fischer, 'Ber.' **23**, 382.
Ethoxysuccinic acid, *C*H.OC$_2$H$_5$.CO$_2$H.CH$_2$CO$_2$H	Penicillium, L-acid	Purdie and Walker, 'Chem. Soc. J. Trans.' 1893, 230.

TRANSLATOR'S NOTE

Fischer and Thierfelder[1] have shown that micro-organisms not only distinguish between isomers of completely opposed activity, but that the transposition of two groups, attached to a single one of a number of asymmetric carbon atoms in a molecule, is of moment to them.

Thus the following sugars are fermentable by various species of yeast:—

$$\text{d-Glucose}\quad CH_2OH\ \begin{matrix} H \\ C \\ OH \end{matrix}\ \begin{matrix} H \\ C \\ OH \end{matrix}\ \begin{matrix} OH \\ C \\ H \end{matrix}\ \begin{matrix} H \\ C \\ OH \end{matrix}\ COH$$

[1] *Ber.* **27**, 2031; see also Fischer, *ibid.* **27**, 2985, 3228, 3479; **28**, 1429.

$$\begin{array}{llccccc} & & H & H & OH & OH & \\ \text{d-Mannose,} & CH_2OH & C & C & C & C & COH \\ & & OH & OH & H & H & \end{array}$$

$$\begin{array}{llccccc} & & H & OH & OH & H & \\ \text{d-Galactose,} & CH_2OH & C & C & C & C & COH \\ & & OH & H & H & OH & \end{array}$$

But the same yeast species are incapable of attacking d-talose,

$$\begin{array}{lccccc} & H & HO & OH & OH & \\ CH_2OH & C & C & C & C & COH \\ & OH & H & H & H & \end{array}$$

which differs from mannose and galactose only by the transposition of the groups attached to a single asymmetric carbon atom.

This result is the more surprising, since changes in the composition of the sugar, though much more marked, do not affect these ferments, which act on sugars with three as well as on those with nine carbon atoms.

Similar results have been obtained with unorganised ferments. To insure fermentation, then, the substance to be fermented and the ferment must have their configurations adjusted to one another like lock and key. It follows that ferments which act on the same substance must resemble one another in configuration like two keys; and they may act on one another to their mutual destruction if the keys turn opposite ways. Experiments made by me showed, however, no such destructive action in the case of human and pig pepsins.

3. SPONTANEOUS DIVISION. TEMPERATURE OF CONVERSION

While the method of division last mentioned depends on the action of the living organism, and the first method is also connected therewith, in that the active compounds employed are mostly products of the organism, the method now to be described does not demand the aid of life. It is a purely chemical one, which isolates the active compound without assistance from animate nature. This method, too, was discovered by Pasteur, who, on crystallising a solution of sodium ammonium racémate, found the two tartrates separated. Although the method has been rarely used since (first by Purdie to divide lactic acid [1]), yet the researches which brought to light the facts on which the method is based have a special interest.

As a matter of history we may remark that Städel,[2] when he evaporated the solution which in Pasteur's hands had yielded the two tartrates, obtained crystals of a double racemate of sodium and ammonium.

This apparent contradiction was harmonised by Scacchi,[3] who showed by a thorough investigation of the racemate in question that a high temperature of crystallisation favours the formation of the racemate, while at the ordinary temperature one obtains

[1] *Trans. Chem. Soc.* 1893, 1143. [2] *Ber.* 11, 1752.

[3] *Rendiconti dell' Accademia di Napoli*, 1865, p. 250.

chiefly the two tartrates. Indeed, Wyrouboff[1] succeeded in showing that the phenomenon is very simple when supersaturation is avoided. There is then a perfectly definite limiting temperature, viz. about 28°; on evaporation one obtains the racemate or the tartrate, according as the crystallisation takes place above or below this temperature.

Temperature of transformation.—The researches which I conducted with van Deventer[2] have shown that we have here to do with a peculiar phenomenon which may occur also outside the solution. The mixture of the two tartrates, when heated a little above 27°, loses a part of its water of crystallisation and is quantitatively converted into the racemate according to the following equation:

$$2C_4O_6H_4NaNH_44H_2O$$
$$= (C_4O_6H_4NaNH_4)_22H_2O + 6H_2O;$$

while below this temperature the reverse takes place. The temperature mentioned is that noticed by Wyrouboff, and the transformation observed gives therefore a complete explanation of his results.

This conversion is also to be detected in the following ways:

1. On mixing the racemate with the above-mentioned proportion of water, the originally soft mass becomes hard, until finally a perfectly dry mixture of the two tartrates remains.

2. A mixture of the two tartrates in equal quantities, heated above 27° in sealed tubes, is partly

[1] *Bull. Soc. Chim.* **41**, 210; **45**, 52; *Compt. Rend.* **102**, 627.

[2] *Zeitschr. f. phys. Chem.* **1**, 173.

liquefied through loss of water of crystallisation and formation of the racemate.

3. The expansion on formation of the racemate renders possible an exact study of the phenomenon. The dilatometer used consisted of a huge thermometer, in the bulb of which was placed the mixture of the two tartrates, this being covered with oil. The height of the oil in the stem of the thermometer was read off on a scale. On heating this dilatometer for a sufficiently long time at definite temperatures, one observes, between 26·7° and 27·7°, a slow but persistent and very considerable expansion, accompanied by a complete change in the contents of the bulb; partial liquefaction takes place, together with production of the racemate in well-formed crystals. On cooling, the reverse phenomenon is observed.

This process of division is of special interest, because it represents the first case of a class of phenomena of chemical equilibrium,[1] much studied of late, and characterised by a definite temperature, the 'conversion temperature,' above and below which only one of the two systems can exist.

Recently van 't Hoff,[2] H. Goldschmidt, and Jorissen have adopted another method for investigating phenomena of this kind. Taking the case already studied, at the temperature of transformation equality will exist in the vapour tension of (1) the

[1] Van 't Hoff and van Deventer, *Ber.* **19**, 2142; *Zeitschr. f. physik. Chem.* **1**, 165, 227; Bakhuis Roozeboom, *Rec. Trav. Chim. Pays-Bas*, **6**, 36, 91, 137; *Zeitschr. f. physik. Chem.* **2**, 336.

[2] *Vorlesungen über Bildung und Spaltung von Doppelsalzen* (Leipzig, 1897, Engelmann).

water of crystallisation of the dextro- and lævo-tartrates, (2) a saturated solution of the above salts, (3) a saturated solution of the salt of Scacchi,

$$(C_4H_4NaNH_4O_6.H_2O)_2.$$

Accordingly the bulbs attached to the two limbs of a differential tensimeter were charged with mixtures representing (1) and (2), and the temperature was observed at which the tension in the two limbs became equal. It was 26·6°.

If the temperature be raised a few degrees another transformation takes place, the double racemate of Scacchi now breaking up to form the single racemates:

$$2(NaNH_4H_4C_4O_6.H_2O)_2$$
$$=(Na_2H_4C_4O_6)_2+((NH_4)_2H_4C_4O_6)_2+4H_2O.$$

At the temperature of conversion there is equality in the vapour tension of

(1) A saturated solution of Scacchi's salt and sodium racemate.

(2) A saturated solution of Scacchi's salt and ammonium racemate.

(3) A saturated solution of sodium and ammonium racemate.

(4) The water of crystallisation of Scacchi's salt.

To represent (4) one division of the tensimeter was filled with 4 gm. of the salt of Scacchi which had been dried till it lost half its water ($\frac{1}{2}$ mol.). Thus was formed the acid ammonium salt which is necessary in order to reduce the ammonia tension to a minimum. The other division contained the same filling, with the addition of one molecule of water;

that is, it was Scacchi's salt *plus* a mixture of dextro- and lævo-tartrates. This was first heated to about 30° so as to form a saturated solution of the salt of Scacchi with one of the two single racemates. The tensions became equal at 34·5°. On account of the presence of the acid ammonium salt this temperature is a little lower than that found by the dilatometer, viz. 36°. It was found, further, that by heating at once above 27° the conversion of the double tartrates to the Scacchi salt could be avoided and the single racemates could be obtained direct. The temperature of this transformation was found to lie, as expected, between the other two, viz. at 29°.

For sodium potassium racemate also, the existence of such a temperature of conversion had been indicated by Wyrouboff's researches.[1] And it is found by van 't Hoff and H. Goldschmidt that at −6° C. this salt is formed from the two tartrates:

$$2NaKH_4C_4O_6.4H_2O = (NaK.H_4C_4O_6.3H_2O)_2 + 2H_2O.$$

Wyrouboff's salt.

This double racemate divides at 41° into the single racemates:

$$2(NaKH_4C_4O_6.3H_2O)_2$$
$$= (Na_2H_4C_4O_6.2H_2O)_2 + (K_2H_4C_4O_6.)_2 + 8H_2O.$$

And at an intermediate temperature, viz. 33°, the direct conversion takes place:

$$4NaKH_4C_4O_6.4H_2O$$
$$= (Na_2H_4C_4O_6.2H_2O)_2 + (K_2H_4C_4O_6.)_2 + 12H_2O.$$

A similar conversion is undergone, *e.g.* by

[1] *Ann. de Chim. et de Phys.* [6], 9, 221.

magnesium sulphate, SO_4Mg7H_2O, and sodium sulphate, $SO_4Na_210H_2O$, these being converted at 21° into a double salt, astracanite, according to the equation:

$$SO_4Mg7H_2O + SO_4Na_210H_2O$$
$$= (SO_4)_2MgNa_24H_2O + 13H_2O.$$

Below 21°, on the other hand, the double salt treated with 13 molecules of water yields the two single salts.

This case, then, is quite analogous to the formation of the racemate from the two constituents, the right- and left-handed tartrates, at 27°.

This third method of division has been employed successfully with racemic and lactic acids. It has been observed [1] also that inactive asparagine, obtained by the action of ammonia on maleïc or fumaric ether, crystallises in hemihedral, enantiomorphous forms, of which the two kinds are present in equal quantity. So also with homoaspartic acid.[2] Also the lactone of gulonic acid [3] divides on crystallising into the two crystals of opposite activity; while an indication of the same thing occurs in the case of dimethyldioxyglutaric acid, but has not yet been utilised for actually dividing it.

More recently Fischer and Beensch have observed a similar transformation in the case of a substance free from water, viz. methyl mannoside, which above 15° exists only in the racemic form, but below 8° only as the two active forms. The

[1] Körner and Menozzi, *Ber.* **21**, Ref. 87.

[2] *Accad. Lincei*, 1893, ii. 368.

[3] Fischer, *Ber.* **25**, 1026.

exact temperature of conversion has not been determined.

4. PROOF OF DIVISIBILITY BY SYNTHESIS OF THE INACTIVE MIXTURE

While in the above cases direct proof of divisibility was afforded by actual division, the observations now to be mentioned are not less convincing. In these, by bringing together two isomers of opposite activity, an inactive body was produced which could then be identified with the inactive product otherwise obtained. Thus the inactive mandelic acid was obtained by Lewkowitsch from the right and left modifications and found to be identical with the synthesised inactive acid, which in fact was afterwards divided. Since then, whole groups of such racemic mixtures have been prepared by Montgolfier, Haller, Jungfleisch, and Friedel: in the camphor series—camphor, borneol, bornylphenylurethane, camphoric acid, &c.; in the terpene series—dipentene and derivatives, limonene, camphene, &c., by Wallach; and finally in the sugar group, by Fischer—arabite, mannite, mannose, glucose, levulose, &c.

Since the result of bringing together two opposed active bodies varies according as this occurs below or above the temperature of transformation—in the one case a mixture resulting, in the other a compound—it is found that, in general, there is a difference in the deportment of active mixtures at a given temperature, say the ordinary temperature. These mixtures belong to two categories. On the one hand an inactive substance is produced which, excepting in optical properties, is exactly similar to the original bodies as regards specific weight, and also chemically; in the other case, however, the product obtained is entirely different from them. Probably the most remarkable examples of this double behaviour are those discovered by Wallach in the terpene series, and by Fischer in the sugar group.

To show the way in which racemic compounds differ from their components, the following list, given by Walden,[1] may be cited.

I. Inactive malic acid, $CO_2H.CH_2.CHOH.CO_2H$ M.P. 130°–131°. Specific gravity, $\frac{20^\circ}{4^\circ}$ (d) = 1·601. Molecular volume, V_m = 83·70. Affinity constant, (K) = 0·040.

L- or natural malic acid. M.P. 100°. d = 1·595. V_m = 84·01. K = 0·040. Solubility greater than the inactive.

II. I-chlorosuccinic acid, $CO_2H.CH_2.CHCl.CO_2H$. M.P. 153°–154°. d = 1·679. V_m=90·83. K=0·294. Solubility (20° C.), 1 in 2·3.

D-chlorosuccinic acid. M.P. 176°. d = 1·687. V_m = 90·40. K = 0·294. Solubility, 1 in 4·5.

L-chlorosuccinic acid. M.P. 176°. d = 1·687. V_m = 90·40. K = 0·294. Solubility, 1 in 4·6.

III. I-bromosuccinic acid, $CO_2H.CH_2.CHBrCO_2H$. M.P. 160°–161°. d=2·073. V_m=95·03. K=0·268. Solubility, 1 in 5·2.

[1] *Ber.* **29**, 1692; compare J. Traube, *l.c.* p. 1394; H. Traube, *l.c.* p. 2446. See also Kipping and Pope on 'Racemism and Pseudo-racemism,' *J. Chem. Soc.* 1897, p. 989.

L-bromosuccinic acid. M.P. 172°. d = 2·093. V_m = 94·12. K = 0·268. Solubility, 1 in 6·3.

IV. I-mandelic acid, $C_6H_5.CHOH.CO_2H$. M.P. 118°–119°. d = 1·300. V_m = 116·9. K = 0·043. Solubility, 15·97 in 100.

L-mandelic acid. M.P. 130°. d = 1·341. V_m =113·3. K = 0·043. Solubility, 8·64 in 100.

V. I-glutamic acid,

$$CO_2H.CH_2.CH_2.CH(NH_2).CO_2H.$$

M.P. 198°. d = 1·511. V_m = 97·29. K, see Walden, 'Z. physik. Chem.' **8**, 489. Solubility, 1 in 59·1.

D-glutamic acid. M.P. 202°. d = 1·538. V_m = 95·58. K, see Walden, *l.c.* Solubility, 1 in 100.

VI. I-camphoric acid, $C_8H_{14}(CO_2H)_2$. M.P. 202°–203°. d = 1·228. V_m = 162·9. K = 0·00229. Solubility, 0·239 in 100.

D-camphoric acid. M.P. 187°. d=1·186. V_m =168·6. K=0·00229. Solubility, 6·96 in 100.

L-camphoric acid. M.P. 187°. d=1·190. V_m =168·1. K=0·00228. Solubility, 6·95 in 100.

VII. I-isocamphoric acid. M.P. 190°–191°. d =1·249. V_m=160·1. K=0·00174. Solubility, 0·203 in 100.

D-isocamphoric acid. M.P. 171°. d=1·243. V_m =160·9. K=0·00174. Solubility, 0·357 in 100.

L-isocamphoric acid. M.P. 171°. d=1·243. V_m =160·9. K=0·00174. Solubility, 0·337 in 100.

VIII. Racemic acid,

$$CO_2H.CHOH.CHOH.CO_2H + H_2O.$$

M.P. 204°. d=1·697. V_m=99·00. K=0·097. Solubility less than the active acids.

D-tartaric acid, $CO_2H.CHOH.CHOHCO_2H$. M.P. 170°. d=1·755. V_m=85·47. K=0·097.

L-tartaric acid. M.P. 170°. d = 1·754. V_m =85·52. K=0·097.

Mesotartaric acid,

$$CO_2H.CHOH.CHOH.CO_2H + H_2O.$$

M.P. 140°. d=1·666. V_m=100·8. K=0·060. More soluble than racemic acid.

As to the melting point, it was to be expected that the mixture should melt lower than its constituents, addition of foreign bodies always lowering the melting point. But from the above examples we see that the racemic form has sometimes a higher, sometimes a lower melting point than its components ; the form of higher melting point being less soluble, and having the smaller molecular volume. (See also Kipping and Pope, 'Proc. Chem. Soc.' 1895, p. 39.)

As to the boiling point, it is similarly to be expected that the compound will have a higher boiling point, but the mixture the same as the constituents, corresponding to the halving of their maximal tension.

If we take this halving of the maximal tension as a basis for calculating the extent to which the melting point is lowered in the cases given above, we have

$$t = 1{\cdot}4\,\frac{T^2}{W},$$

where T is the absolute melting temperature, W the latent heat of liquefaction per kilogram-molecule. In fact this equation gives values which fairly correspond with the observations in the case of the gulonic acid lactones. Taking the known heat of liquefaction for organic compounds,[1] we get numbers between 15° and 45°. This could be exactly tested by determining the heat of liquefaction of the gulonic lactones, and so perhaps we should arrive at a new means of determining racemic character.

Proof of divisibility without direct division.—In the above cases it was possible to prove divisibility indirectly, by synthesis of the inactive compounds; but even if only one of the products of division is known, we may obtain the desired proof, in the case of acids at least, by examining the conductivity. Since the divisible compounds, such as racemic acid, are decomposed in solution,[2] and the conductivity of the two active components is the same, it is sufficient to prove that the conductivity of the inactive body is equal to that of the active one. As is well known, the electrical deportment varies so strongly with the constitution, that identity in one respect makes identity in the other extremely probable.

That inactive malic acid[3] showed in Ostwald's research the same conductivity as the active acid, convinces us therefore of the divisibility of the former. The same conclusion is to be drawn from Eykman's[4] examination of inactive quinic acid, which also was found to be equal to the active acid.

The peculiar behaviour of compounds containing

[1] Eykman, *Zeitschr. physik. Chem.* **3**, 209. [2] Raoult, *ibid.* 371.
[3] *Zeitschr. physik. Chem.* **3**, 370. [4] *Ber.* **24**, 1289.

an asymmetric carbon atom, resulting from symmetrical bodies in ordinary laboratory experiments, is now explained, and it only remains to mention that, in what may be called asymmetric conditions of formation, we may expect another and a simpler state of things, and that we find it. The speed of formation of the two isomers is in this case generally different and the product directly active. This is illustrated by the direct formation of active bodies in the organism, an apparatus consisting essentially of active materials; thus, from inactive carbonic acid, water, ammonia, and nitrates, the plant forms the innumerable active compounds with which we are familiar—terpenes, carbohydrates, alkaloids. In the animal organism, which consumes principally active material, the opportunity for such observations is evidently more limited; yet Baumann and Preusse[1] were able to show that inactive bromobenzene is converted in the body into bromophenylmercapturic acid.[2]

It is exceedingly probable that in other asymmetric conditions of experiment the same direct formation of active bodies will result; *e.g.* in transformations taking place under the action of right or left circular polarised light, or caused by active com-

[1] *Zeitschr. physiol. Chem.* **5**, 309; *Ber.* **15**, 1731.

[2] The production of the active acid may, however, be due to preliminary formation of active alanine, and therefore cannot be taken as proof of the production of active from inactive compounds by the animal organism (*J. f. physiol. Chem.* **21**, 255). On the other hand it has been observed that, after poisoning by carbon monoxide, the injection of inactive sodium lactate produces separation of active acid (*l.c.* **19**, 455; **20**, 374).

pounds, perhaps even if only taking place in active solvents.[1]

Indivisibility when the asymmetric carbon atom is absent.—If on the one hand it may be said that up till now no compound with an asymmetric carbon atom, when suitably treated, has escaped division (with the exception of certain compounds of symmetrical type which will be considered later), it is important on the other hand to observe that division has been repeatedly attempted in the absence of dissymmetry, but so far without success.

We may again mention here the compounds enumerated on page 23, obtained from active bodies by fermentation &c., and yet inactive; we may add the inactive vegetable products. Of special importance, however, are the experiments undertaken with the express purpose of obtaining division: *e.g.* of oxalic acid by Anschütz and Hintze;[2] of fumaric acid[3] by the same; of orthotoluidine[4] by Le Bel; of inosite[5] by Maquenne; of homosalicylic acid, $C_6H_3(CH_3)(CO_2H)OH(1, 2, 3)$, of homo-oxybenzoic acid, $C_6H_3(OH)(CH_3)CO_2H(1, 2, 3)$, and of methoxy-

[1] Pope and Kipping (*Proc. Chem. Soc.* December 1896) find that substances active only in the solid form, which ordinarily deposit from solution dextro- and lævo-gyrate crystals in equal numbers, may be made to yield an excess of one form by dissolving an active substance with them. Thus 5 per cent. of dextrose in the solution caused a preponderance of l-sodium chlorate in the separated crystals, while 5 per cent. of isodulcite caused the dextro-chlorate to preponderate. They suggest that racemic bodies may be divided in this way. (Compare Eakle, *Chem. Centralbl.* 1896, ii. 649.)

[2] *Ber.* **18**, 1394.

[3] *Ann.* **239**, 164.

[4] *Bull. Soc. Chim.* **38**, 98.

[5] *Compt. Rend.* **104**, 225.

toluylic acid, $C_6H_3(OCH_3)(CH_3)CO_2H$(1, 2, 3), by Lewkowitsch.[1] All these attempts were unsuccessful.

Mutual conversion of active bodies. Position of equilibrium.—From the exactly corresponding configuration of the two isomers, it is at once clear that the stability of both is the same. Now this stability is, in general, slight; it has long been known that on warming sufficiently the activity is lost while the composition remains the same. It has gradually become certain that we have here to do with the formation of the racemic mixture. On heating tartaric acid, racemic acid[2] is formed; on heating active amylalcohol (as sodium derivative) the result is an inactive product, which Le Bel[3] has divided; mandelic acid yields an inactive mixture, divided by Lewkowitsch;[4] Schulze and Bosshard[5] obtained by heating active leucine an inactive isomer, which was also divided by them; Michael and Wing[6] obtained aspartic acid, divided by Engel;[7] Wallach the divisible dipentene from active isomers. In short, we have here a general method for preparing from one isomer a derivative of opposed activity—viz. racemising by heat and dividing the product. Further, Walden[8] has made the curious observation that lævomalic acid produces with PCl_5 the

[1] *J. Chem. Soc. Trans.* 1888, 781.

[2] Jungfleisch, *Compt. Rend.* **75**, 439, 1739.

[3] *Bull. Soc. Chim.* **31**, 104; *Compt. Rend.* **87**, 213.

[4] *Ber.* **15**, 1505.

[5] *Ibid.* **18**, 588; *Zeitschr. physiol. Chem.* **10**, 134.

[6] *Ber.* **18**, 2984.

[7] *Compt. Rend.* **106**, 1734.

[8] *Ber.* **28**, 2766; **29**, 133.

chlorosuccinic acid corresponding to dextromalic acid; ethyl lactate behaves in the same way.[1] Similarly Anschütz, on treating fumaric acid with bromine, observed the formation of the dibromosuccinic acid corresponding to inactive tartaric acid; whereas, according to page 107 *post*, the racemic compound was to have been expected.

It is important to add that so-called catalytic influences may bring about a similar racemising, as in the conversion of hyoscyamine into atropine by bases,[2] and of tartaric into racemic acid by the oxides of iron and aluminium;[3] in fact these catalytic influences occasionally cause the formation of racemic acid in the commercial preparation of tartaric acid.

Further, the isomerisation in question, leading to the production of inactive bodies, takes place more easily during their formation than with the ready-formed bodies, which is in perfect accord with all our conceptions of the nascent state. Thus on heating albuminoids with baryta, the result was inactive tyrosine, leucine, and inactive glutamic acid, while when Schulze and Bosshard used hydrochloric acid for the conversion, all the derivatives were obtained in the active state; in the former case, then, isomerisation with loss of activity was brought about by the united action of heat, alkali, and the nascent condition. In those cases which have not yet been cleared up by direct experiment there is yet hardly room for doubt. That nitro- and pyro-tartaric acid

[1] Purdie, *J. Chem. Soc.* 1896, p. 818.

[2] *Ber.* **21**, 2777.

[3] Jungfleisch, *Compt. Rend.* **85**, 805.

are inactive, although asymmetrical and prepared from tartaric acid, is probably due to the same cause. Hence also arises the inactivity of bromosuccinic acid made from malic acid; it has already been mentioned on p. 24 that in the corresponding reaction the chlorine product was obtained active. It should be specially mentioned here that, when halogens are brought into union with the asymmetric carbon, isomerisation takes place very readily, as is shown by the substances mentioned on p. 24—dichlorosuccinic acid, iodohexyl, phenylchlor-, brom-, and isopropylphenylchlor-acetic acid.

The case is of course quite otherwise with Hartmann's[1] anhydride of active camphoric acid, which changes back into the active camphoric acid; it chanced to be inactive under the particular conditions, as, according to Colson,[2] may be the case with the isobutylamyl ester.

Finally, it is of special interest to consider the matter from a more general kinetic and thermodynamic point of view. The state of equilibrium of active isomers in a racemic mixture is the simplest conceivable. Kinetically it is evident that, if the stability is slight and leads to conversion, equilibrium will be attained when the inactive mixture is formed. Since, from the complete mechanical symmetry, the tendency to conversion is equal in the two isomers, the one present in larger quantity will always be converted in larger quantity, until equal quantities of each are present.[3]

[1] *Ber.* 1888, 221. [2] *Compt. Rend.* February 1893.
[3] Van 't Hoff, *Ber.* **10**, 1620.

Thermodynamically we have here a case most remarkable for its simplicity. Seeing that the equilibrium depends upon the work, E, which can be performed by the conversion, and which in our case, owing to the mechanical symmetry, is evidently *nil*, the equilibrium-constant K, which determines the proportion of the two active substances, must be unity, according to the equation :

$$log\ K = -\frac{E}{2T},$$

where T is the absolute temperature.[1]

A word in parenthesis. We have here one of those rare cases in which alteration of the temperature does not change the equilibrium-constant ; and this is simply because this change depends on the heat of conversion, q (which is here *nil*—again owing to the mechanical symmetry), according to the equation :

$$\frac{d\ log\ K}{dT} = \frac{q}{2T^2}.$$

This is the simplest form of equilibrium.

INACTIVE, INDIVISIBLE TYPE

When Le Bel's paper and mine were laid before the Société Chimique in Paris, Berthelot[2] observed that our views took no account of the 'indivisible nactive type.' It was Pasteur who described this

[1] Van 't Hoff, *Arch. Neerl.* 1886 ; *Kongl. Svenska Akad. Handl.* 1886.

[2] *Études de dynamique chimique.*

modification in the case of tartaric acid, an indivisible inactive tartaric acid being known, as representative of the fourth type, in addition to the two active acids and the combination of these two. In fact, in this special case, in the presence of two asymmetric carbon atoms our theory foresees the existence of an indivisible inactive compound.

Since then some chemists have assumed the existence of this indivisible inactive modification as quite general. This type is, indeed, not to be explained by the theory in cases where the constitutional formula contains only one asymmetric carbon atom. In this respect the objection of Berthelot was perfectly justified. The next thing was to find a representative of the inactive type in bodies containing only a single asymmetric carbon atom, and Berthelot instanced the inactive malic acid, which was, indeed, the only compound presenting a serious objection to our theory. This malic acid had been obtained by Pasteur from the inactive aspartic acid of Dessaignes. This malic acid was inactive, and Pasteur mentions it as indivisible.[1] But his paper did not give me the impression that he wished to bind himself very strongly to this statement.

Nevertheless, the existence of the indivisible inactive malic acid, in addition to the inactive compound resulting from compensation, had from that time been generally accepted.[2]

[1] *Ann. de Chim. et de Phys.* [3], **34**, 46.

[2] See, *e.g.* Landolt, *Das optische Drehungsvermögen organischer Substanzen*, p. 20.

Through the more recent researches of Bremer, Anschütz, and H. J. van 't Hoff, this difficulty has now been removed. Not only has Pasteur's acid been studied afresh, but all inactive malic acids, prepared according to the methods at present known, were identified with the inactive acid which results from mixing equal quantities of right- and left-handed acids; while more than one of these acids was found capable of division. My brother[1] proved the identity of the synthesised inactive acid with that of Pasteur. He prepared its acid ammonia salt in the two forms which are observed in the case of Pasteur's acid, according as the salt is anhydrous or hydrated. He proved[2] the same thing for the inactive acid which had been obtained by Loydl by heating fumaric acid with caustic soda, and this identity was confirmed by the division effected by Bremer.[3] The same crystalline form was observed in Kekulé's acid ammonium malate prepared from bromosuccinic acid.

Anschütz[4] found on a crystallographic comparison of these salts, made from the acids of Pasteur, Kekulé, and Jungfleisch (the last obtained by heating fumaric acid with water), that here also identity exists. Jungfleisch,[5] too, found his acid ammonium malate to be crystallographically identical with that of Pasteur, though he does not give the measurements.

[1] *Maandblad voor Natuurwetenschappen*, 1885, *Bijdrage to de kennis der inaktieve Appelzuren*. Diss. 1885.

[2] *Ber.* **18**, 2170.

[3] *Rec. Trav. Chim. Pays-Bas*, **4**, 180.

[4] *Ber.* **18**, 1949.

[5] *Bull. Soc. Chim.* **30**, 147.

This evidence has been strengthened by an observation of Piutti,[1] who recognised in the inactive aspartic acid, obtained by mixing the right and left isomers, the acid prepared by Dessaignes, which was the one used by Pasteur for making malic acid. We may add that the malic acid[2] prepared by heating maleïc acid with caustic soda has been found identical with the divisible màlic acid, an observation which has since been confirmed by the probable identity of the methyl- and ethyl-malic acids which Purdie[3] obtained on treating fumaric and maleïc acids with sodium methylate and ethylate. Thus the isomerism of these acids disappears in the malic acid derivatives formed from them.

Then, too, what Fischer observed in the case of galactonic acid is most important. From mucic acid, which in accordance with its symmetrical formula, $CO_2H(\mathit{C}HOH)_4CO_2H$, can be inactive and indivisible, as indeed it is, he obtained a galactonic acid, $CO_2H(\mathit{C}HOH)_4CH_2OH$, *not* of the inactive indivisible type, but a product which was capable of division; surely a proof that even under the most favourable conditions no inactive indivisible type results, unless its existence is justified by the symmetrical constitution.

[1] *Compt. Rend.* **103**, 134.
[2] *Ber.* **18**, 2173.
[3] *Chem. Soc. J.* 1885, 855.

CHAPTER III

COMPOUNDS WITH SEVERAL ASYMMETRIC CARBON ATOMS

I. Application of the Fundamental Conception

Spatial arrangement. Free rotation.—From the hypothesis that the groups attached to an asymmetric carbon atom correspond to an unsymmetric tetrahedron, it follows at once that for a compound with two asymmetric carbon atoms joined to each other, $CR_1R_2R_3$ $CR_4R_5R_6$, the arrangement is determined as far as this: each carbon atom must form at once the centre of one and the corner of the other tetrahedron, as shown in the accompanying figure.

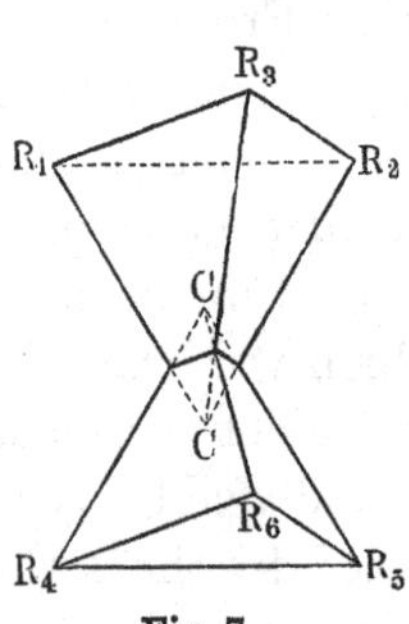

Fig. 7.

Every other arrangement, however, obtained by rotation (of the lower tetrahedron, *e.g.*) round the axis C—C must be equally in harmony with the fundamental conception. But in order to avoid this idea of an infinite isomerism, no additional hypothesis is necessary. Free rotation being admitted by the fundamental conception, the mutual action of the groups R_1 R_2 R_3 on the one hand, and R_4 R_5 R_6 on the other will lead to a single 'favoured configuration.' It is for the present indifferent which we call the 'favoured configuration,' and we may take as such the arrangement represented by the figure, where R_1 is above R_4, R_2 above R_5, R_3 above R_6.

We may use now either the model recommended on p. 8, or, following Friedländer, we may improvise one from caoutchouc tubes and sealing-wax, each tetrahedral grouping being represented by four short tubes meeting at equal angles.[1] Or, instead of models, we may, as has been proposed by Fischer and also by myself, choose as the most suitable way of representing these isomers, a projection, in which the front groups R_3 and R_6 are turned upwards or downwards, and so appear on paper thus:

$$\begin{array}{ccc} & R_3 & \\ R_1 & C & R_2 \\ R_4 & C & R_5 \\ & R_6 & \end{array}$$

These projections may conveniently be used for illustrating the possible isomers. That there are four of these is at once evident, since each asymmetric carbon involves a doubling.

These differences are represented by changing the order of the groups R_1 R_2 R_3. But if, without

[1] It is convenient to attach the tubes to a hollow caoutchouc ball (Jowett's model).

changing this order, we simply move R_1 to R_2, R_2 to R_3, R_3 to R_4, we only bring about the above-mentioned rotation and no isomerism results. But if R_1 and R_2 change places we get a new isomer; also by transposing R_4 and R_3; hence the following symbols represent the four isomers:

No. 1	No. 2	No. 3	No. 4
R_3	R_3	R_3	R_3
R_1 C R_2	R_2 C R_1	R_1 C R_2	R_2 C R_1
R_4 C R_5	R_4 C R_5	R_5 C R_4	R_5 C R_4
R_6	R_6	R_6	R_6

It is plain that these are reduced to two directly the asymmetry of one of the carbon atoms ceases through R_5 and R_4 becoming identical; the difference between No. 1 and No. 3 on the one hand, and between No. 2 and No. 4 on the other, then vanishes.

Several asymmetric carbon atoms.—The special advantage of these projections lies in the fact that they may almost be said to gain in simplicity when the cases become more complicated. For representing the isomerism in the simplest cases they are not well adapted; here they have to compete with the cardboard model. For three asymmetric carbons they are decidedly superior. The number being then again doubled we have to expect eight isomers; in general, 2^n for n asymmetric carbons. The middle carbon atom then holds two groups:

$$CR_1R_2R_3CR_7R_8CR_4R_5R_6.$$

If now the configurations are worked out in three dimensions—according to Friedländer, *e.g.*—and then utilising the free rotation, one of the simplest possible positions is chosen, where R_1 R_2, R_7 R_8, and R_4 R_5 are in parallel lines, the projection leads to a formula like that below, and the eight isomers result on simply transposing R_1 and R_2, R_7 and R_8, R_4 and R_5 :

$$
\begin{array}{ccccccccccccccc}
 & \text{No. 1} & & & & \text{No. 2} & & & & \text{No. 3} & & & & \text{No. 4} & \\
 & R_3 & & & & R_3 & & & & R_3 & & & & R_3 & \\
R_1 & C & R_2 & & R_2 & C & R_1 & & R_1 & C & R_2 & & R_2 & C & R_1 \\
R_7 & C & R_8 & & R_7 & C & R_8 & & R_7 & C & R_8 & & R_7 & C & R_8 \\
R_4 & C & R_5 & & R_4 & C & R_5 & & R_5 & C & R_4 & & R_5 & C & R_4 \\
 & R_6 & & & & R_6 & & & & R_6 & & & & R_6 & \\
 & \text{No. 5} & & & & \text{No. 6} & & & & \text{No. 7} & & & & \text{No. 8} & \\
 & R_3 & & & & R_3 & & & & R_3 & & & & R_3 & \\
R_1 & C & R_2 & & R_2 & C & R_1 & & R_1 & C & R_2 & & R_2 & C & R_1 \\
R_8 & C & R_7 & & R_8 & C & R_7 & & R_8 & C & R_7 & & R_8 & C & R_7 \\
R_4 & C & R_5 & & R_4 & C & R_5 & & R_5 & C & R_4 & & R_5 & C & R_4 \\
 & R_6 & & & & R_6 & & & & R_6 & & & & R_6 &
\end{array}
$$

II. Experimental Confirmation

A. NUMBER AND CHARACTER OF THE ISOMERS TO BE EXPECTED

The actual preparation of the isomers (which, as we have seen, should number four, eight, or sixteen, according as two, three, or four asymmetric carbon atoms are present) is simple, provided this can be effected by combining several compounds, each containing one asymmetric carbon. Let us take, for instance, as a starting point, the two active malic acids; with these and the two active amylalcohols we can evidently prepare four amylmalic acids. If now we introduce in place of the hydroxyl hydrogen atom the radical of any acid obtainable in two forms of opposite activity, *e.g.* lactic acid, we should have the eight bodies, which, on saturating with the two conines of opposite activity, would give us the required number of sixteen.

As to the characters of the isomers, a consideration of the above, as well as a glance at the formulæ Nos. 1 to 4 and Nos. 5 to 8, shows that the isomers would be grouped in pairs. The symbols No. 1 and No. 4, No. 2 and No. 3, in the first series, and those which may easily be found in the second series, are reflections one of the other, and stand to each other therefore as the two isomers with one asymmetric carbon atom, and are, like them, so similar that one might be taken for the other. The existence of the second asymmetric carbon atom only betrays itself by the appearance of a second type, which is also present in two forms, but in general differs from the first type in activity, melting point, and solubility. For example, in the case of the experiment which we have supposed to be carried out, No. 1 and No. 4 would correspond to left-amyl left-malic acid and right-right acid; No. 2 and No. 3 to right-amyl left acid and left-amyl right acid.

The following have been obtained:

Two asymmetric carbon atoms.—The four active borneols and their derivatives, obtained by Montgolfier[1] and Haller[2] by reduction of camphor, exactly correspond with the above.

[1] *Thèses sur les Isomères et les Dérivés du Camphre et du Borneol.*
[2] *Compt. Rend.* **105**, 227; **109**, 187; **110**, 149; **112**, 143.

Camphor	Borneol
C_3H_7	C_3H_7
C	C
HC CH_2	HC CH_2
H_2C CO	H_2C *C*H(OH)
C	*C*
HCH_3	HCH_3

r- and l-Borneol (*a*) $[a]_D$ = + and − 37° (in alcohol), + and − 38° (in toluene)

r- and l-Borneol (*β*) $[a]_D$ = + and − 33° (in alcohol), + and − 19° (in toluene)

Borneolphenylurethane (*a*) $[a]_D$ = + and − 35° (M.P. 137°)

Borneolphenylurethane (*β*) $[a]_D$ = + and − 57° (M.P. 130°)

A second example is afforded by limonenenitrosochloride and its derivatives:[1]

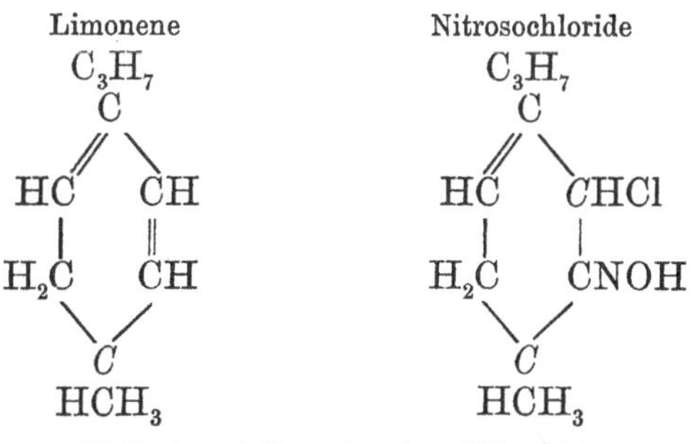

[1] Wallach and Conrady, *Ann.* **252**, 144.

r- and l-*α*-Nitrosochloride	$[\alpha]_D$ = + and − 313° (M.P. 103°)
r- and l-*β*-Nitrosochloride	$[\alpha]_D$ = + and − 241°
r- and l-*α*-Nitrolepiperidine	$[\alpha]_D$ = + and − 68° (M.P. 94°)
r- and l-*β*-Nitrolepiperidine	$[\alpha]_D$ = + and − 60° (M.P. 110°)

Perhaps the four camphoric acids come in the same category:

$$\begin{array}{ll} HC{=}C(C_3H_7)CO_2H & CH_2\textit{C}(CH_3)CO_2H \\ \quad | & | \quad\;\; | \\ H_2C\textit{C}H(CH_3)CO_2H & CH_2\textit{C}(C_3H_7)CO_2H \\ \text{(Kekulé [1])} & \text{(Bamberger [2])} \end{array}$$

r- and l-Camphoric acid (*α*)	$[\alpha]_D$ = + and − 46° (M.P. 180°), dissolves in 160 parts of water at 15°
r- and l-Camphoric acid (*β*) [3]	$[\alpha]_D$ = + and − 46° (M.P. 113°), dissolves in 268 parts of water at 15°

The constitution being here uncertain, however, the possibility of an isomerism like that of fumaric and maleïc acids is not excluded, and this would explain the remarkable equality in the rotating power of the *α* and *β* modifications. This occurs, however, in other cases and will be discussed later.

[1] *Ber.* **6**, 932. [2] *Ibid.* **23**, 218.

[3] Friedel, *Compt. Rend.* **108**, 978; Jungfleisch, *l.c.* **110**, 790; Marsh, *Ber.* **23**, Ref. 229.

In the case of atropine, which, from its decomposition products, tropine and tropic acid,[1] contains at least two asymmetric carbons:

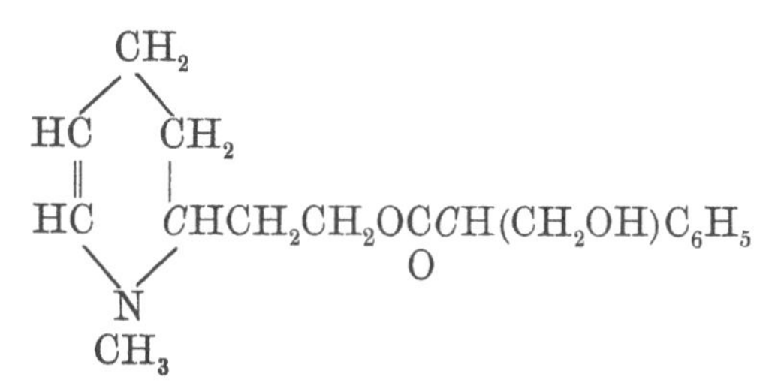

there have been obtained the left- and right-rotating atropines,[2] $[\alpha]_D = 10°$, M.P. 110°; of the two other modifications only one is known, viz. hyoscyamine, $[\alpha]_D = -21°$, M.P. 109°. The atropine formed from this by the action of bases, which according to Hesse[3] is active, may belong to a third type.[4]

There is still another point on which more light is required. The left-handed atropine, resulting from left-atropic acid and inactive tropine, could also exist in several modifications, since this tropine must be considered as a racemic mixture.

In the case of the allied substance, cocaïne ($[\alpha]_D = -16°$;[5] as HCl salt $[\alpha]_D = -68°$), and of ecgonine (as HCl salt $[\alpha]_D = -62°$[6]):

[1] Ladenburg, *Ber.* **21**, 3065; Einhorn, *l.c.* **23**, 1338.

[2] Ladenburg, *Ber.* **22**, 2591.

[3] *Ann. Chem.* **271**, 100.

[4] Pseudohyoscyamine, $[\alpha]_D = -21°$, gives on decomposition an isomeric tropine, like hyoscine (*J. Chem. Soc. Abstr.* 1893, p. 491).

[5] *Ber.* **20**, 320.

[6] Einhorn, *Ber.* **22**, 1495.

Cocaïne

$$
\begin{array}{l}
\quad CH_2 \\
HC \quad\; CH_2 \\
\| \qquad\; | \\
HC \quad\; CH\mathit{C}H(OC_7H_5O)CH_2CO_2CH_3 \\
\quad\; N \\
\quad CH_3
\end{array}
$$

Ecgonine

$$
\begin{array}{l}
\quad CH_2 \\
HC \quad\; CH_2 \\
\| \qquad\; | \\
HC \quad\; CH\mathit{C}H(OH)CH_2CO_2H \\
\quad\; N \\
\quad CH_3
\end{array}
$$

the transformation which the latter undergoes when treated with alkalies corresponds to that undergone by hyoscyamine, only there results here a right ecgonine[1] (as HCl salt $[a]_D = +20°$) which yields a right cocaïne (as HCl salt $[a]_D = +42°$). Both are therefore representatives of the β type.

However, here too the constitution is doubtful. If we were to accept Merling's[2] tropine formula:

$$
\begin{array}{l}
\quad CH_2 \\
HC \quad CH_2 \\
\| \qquad | \\
HC \quad \mathit{C}HCH_2CH_2OH \\
\quad N \\
\quad CH_3
\end{array}
$$

(Ladenburg)

$$
\begin{array}{l}
\qquad\quad CH_2 \\
HC \qquad\qquad CH_2 \\
\quad\; CH_2 \\
\qquad\qquad CHOH \\
H_2C \qquad\qquad \mathit{C}H \\
\qquad\quad N \\
\qquad\quad CH_3
\end{array}
$$

(Merling)

[1] Einhorn, *Ber.* **23**, 468; Liebermann, *l.c.* 511.

[2] Liebermann, *Ber.* **25**, 929.

we should expect to find four asymmetric carbons in atropine and cocaïne, and accordingly sixteen isomers.[1] And Merling's formula has been rendered the more probable, since Willstätter has found two inactive tropines (probably racemic). Ladenburg's formula admits of only one.

Three asymmetric carbon atoms.—Here there should be eight isomers, in four pairs. If we take the left and right tartaric acids as corresponding to substances with a single asymmetric carbon atom, then we find that a not inconsiderable proportion of these eight isomers has been prepared by Wallach and Conrady [2] from limonene :

r- and l-α-Nitrolebenzylamine right tartrate	$[\alpha]_D - 50°$ and $+ 70°$
r- and l-α-Nitrolebenzylamine left tartrate	$[\alpha]_D - 70°$ and $+ 50°$

but the four isomers of the β type are lacking.

Also, Haller has obtained chloralbornylates,

$$CCl_3.\mathit{C}H(OH)C_{10}H_{17}O,$$

in four modifications, r- and l- (α), and r- and l- (β) ; while it is quite possible that in each of the four methods of preparation, *e.g.* from chloral and r-borneol (α), two isomers were formed.

In the sugar series, in the pentose group,

$$COH(\mathit{C}HOH)_3CH_2OH,$$

[1] *Ber.* **29**, 936.

[2] *Ann. Chem.* **252**, 144.

we have:

r- and l-Arabinose[1] $[\alpha]_D = +$ and $- 104°$ (at first 157°)
Xylose $[\alpha]_D = + 19°$ (at first 79°)
Ribose[2]

Here, then, three types have been experimentally realised, which also have been found to recur in the corresponding acids:

Arabonic acid $[\alpha]_D < - 8°$; as lactone $[\alpha]_D = - 74°$
Xylonic „ $[\alpha]_D = - 7°$; „ „ $[\alpha]_D = + 21°$
Ribonic „ ? „ „ $[\alpha]_D = - 18°$

Four asymmetric carbon atoms.—Of the sixteen isomers (eight types) there have been prepared, in the case of the glucoses,

$$COH(\mathit{C}HOH)_4CH_2OH,$$

and particularly by Fischer:[3]

l- and r-Glucose $[\alpha]_D = 53°$ (at first 105°)
l- and r-Mannose[4] $[\alpha]_D = 13°$
l- and r-Gulose
Galactose $[\alpha]_D = 80°$ (at first 118°)
Talose
Idose

Of the corresponding acids,

$$CO_2H(\mathit{C}HOH)_4CH_2OH,$$

and their lactones, there have been obtained:

l- and r-Gluconic acid $[\alpha]_D = 8°$; as lactone, 68°
l- and r-Mannonic acid $[\alpha]_D = 3°$; „ „ 54°

[1] *Ber.* **26**, 740.
[2] Fischer, *Ber.* **24**, 4220.
[3] *Ber.* **24**, 1840, 3622.
[4] *l.c.* **22**, 368, 3218.

l- and r-Gulonic acid [1] $[\alpha]_D = 14°$; as lactone, 55°
Galactonic acid $[\alpha]_D < -11$; " " $-71°$
Talonic acid,[2] as lactone, strongly left-handed.

B. FORMATION OF THE ISOMERS WITH SEVERAL ASYMMETRIC CARBON ATOMS

Whereas the formation of a compound with several asymmetric carbon atoms by the union of two substances, each containing one such atom, leads to results which can easily be foreseen, the case is altered when the number of the asymmetric carbon atoms increases through a transformation taking place within the molecule.

First case.—Theoretically the simplest case is that in which we start from a single compound with an asymmetric carbon atom, that is, a compound active and not racemic.[3] If we introduce into such a compound a new asymmetric carbon atom, as in the transformation of camphor to borneol, we may in general expect the production of two isomers. But this case is quite distinct from that in which the original compound contained no asymmetric carbon atom, where the resulting isomers are images of each other and possess that identity of internal structure which causes the formation of equal quantities of each. Here the case is different. We have now to do with conditions like those which

[1] *l.c.* **24**, 526. [2] *l.c.* **24**, 3625.

[3] To indicate a compound rendered inactive by the mutual counterbalancing of two active isomers, it is as well to use the word chosen by Pasteur, with whom this conception originated.

determine, *e.g.* the formation of the right malate of right and of left conine. The formation of equal quantities of the two isomers, which will in general possess unequal stability, is by no means to be predicted; indeed, one isomer may predominate to such a degree that the other escapes detection; further, the two isomers may be separated by ordinary means, *e.g.* by crystallisation, without necessitating a resort to the special means of dividing optical isomers.

The researches of Montgolfier and Haller mentioned above afford the most suitable illustration of all this. On conversion into borneol, camphor gives two isomers, the right-handed, stable, ordinary modification (α), and a left-handed unstable modification (β). These may be separated from each other by simple crystallisation, and yield on oxidation the same original camphor. The left-handed matricaria camphor also forms two complementary compounds, as shown in the following table:

Ordinary camphor	right stable borneol	$[\alpha]_D = +37°$
$\alpha_D = +55°$	left unstable „	$[\alpha]_D = -33°$
Left camphor	left stable „	$[\alpha]_D = -37°$
$\alpha_D = -55°$	right unstable „	$[\alpha]_D = +33°$

We may add that turpentine oil also yields two isomeric borneols[1] on treatment with sulphuric acid, and that there are two camphoric acids corresponding to each camphor.

[1] Bouchardat and Lafont, *Compt. Rend.* **105**, 49.

Similarly, E. Fischer,[1] by addition of hydrocyanic acid, has formed two acids, l-mannonic and l-gluconic acids, from arabinose; *α*- and *β*-gluco-heptonic acid from glucose; and *α*- and *β*-gluco-octonic acid from heptose, the asymmetric group, $XCH(OH)CO_2H$, being introduced in place of the aldehyde group. Glucose yields, moreover, two isomeric methylglucosides.[2] Finally, by reduction of levulose: $CH_2OH(CHOH)_3COCH_2OH$, asymmetry is introduced into the CO group, and the result is the simultaneous formation of two isomeric alcohols, mannite and sorbite.[3]

There is every reason to class in this category the formation of the isomeric nitrosochlorides which were obtained by Wallach[4] from limonene (p. 60), by means of amylnitrite and hydrochloric acid; the right limonene gives, *e.g.* an *α*- and *β*-nitrosyl-chloride (chlorinated oxime). It seems indeed, at the first glance, somewhat strange that each of these chlorides should yield with amines—*e.g.* piperidine—a mixture of *α*- and *β*-nitrosamine; but this is probably due to an isomerisation taking place during the transformation, such as, according to p. 49, is especially apt to occur in the case of halogen derivatives.

Finally, in the fact that left and right ecgonine

[1] *Ber.* **23**, 2611; **24**, 2685; *Ann.* **270**, 64.

[2] Alberda v. Ekenstein, *Rec. des Trav. Chim. des Pays-Bas*, 1894, p. 183.

[3] *Ber.* **23**, 3684; Meunier, *Compt. Rend.* **111**, 49.

[4] *Ann.* **252**, 106; *Ber.* **23**, 3687; **24**, 1653, 2687.

(p. 62) yield the same active anhydrecgonine and the same ecgonic and tropic acid, we may see another example of the disappearance of isomerism consequent on the elimination of an asymmetric carbon atom.[1]

Second case.—A case theoretically more complicated, but often realised in the laboratory, occurs when two asymmetric carbon atoms are introduced into an inactive or racemic compound, as in the addition of bromine to cinnamic acid, forming $C_6H_5(\mathit{C}HBr)_2CO_2H$, and in the addition of nitrosylchloride to dipentene (= racemic limonene). In both cases we have to expect the formation of an inactive mixture, consisting of two racemic pairs, represented by:

First pair: $+A+B$ and $-A-B$
Second pair: $+A-B$ and $-A+B$

The ordinary methods of separation yield, then, two (racemic) isomers, the special methods yielding four.

It is only lately that such cases have been experimentally demonstrated. Wallach was able to follow them out in detail by preparing from l- and r-limonene-α-nitrosylchloride the inactive, or i-nitrosylchloride (α), and then in the same way the i-nitrosylchloride (β). On treating dipentene he then obtained and isolated the i-(α)- and i-(β)-products.

Erlenmeyer, jun., Lothar Meyer, jun., and Liebermann have broken up the cinnamic acid bromide ($\alpha_D=68°$) and the last named the dichloride also

[1] *Ber.* **24**, 611.

(a_D=67°). The division of bromophenyl-lactic acid is the third example of division in presence of two asymmetric carbon atoms. But the third and fourth isomers are still lacking. On the other hand, there is a whole series of as yet undivided racemic compounds, which have already been obtained in the two isomers foreseen by the theory. Such are : bromine addition products of crotonic and isocrotonic acids, $CH_2(\mathit{C}HBr)_2CO_2H$,[1] angelic and tiglic acids, $CH_3\mathit{C}HBr\mathit{C}BrCH_3CO_2H$,[2] hypogæic and gaidinic acids, $CH_3(\mathit{C}HBr)_2C_{13}H_{25}O_2$, oleic and elaidic acids, $CH_3(\mathit{C}HBr)_2C_{15}H_{29}O_2$, erucic and brassidic acids, $CH_3(\mathit{C}HBr)_2C_{19}H_{37}O_2$, mesaconic and citraconic acids, $CH_3\mathit{C}Br(CO_2H)\mathit{C}HBrCO_2H$. Such also are the bi-substituted succinic acids, all of which have been obtained in two modifications : brommethyl-,[3] methylallyl-, allylethyl-, benzylmethyl-, benzylethyl-, methylphenyl-succinic acid[4]; further, methylethyl and methylpropyl-glutaric acids, and isomeric glycols, $X(CHOH)_2Y$, like phenylmethylglycol,[5] which, according to Zincke,[6] constantly occur in two modifications.

Interesting, too, is the fact lately established by Schiff,[7] that crotonchloral gives with amides (acet-

[1] Melikoff, *Ber.* **16**, 1268; Wislicenus, **20**, 1010.
[2] Pückert, *Ann.* **250**, 244; Fittig, **259**, 1.
[3] Bischoff, *Ber.* **23**, 3622.
[4] *Ibid.* **24**, 1876; *Zeitschr. f. physiol. Chem.* **8**, 465.
[5] Zincke, *Ber.* **17**, 708.
[6] *Ber.* **20**, 339.
[7] *Ibid.* **26**, 446; see also Griner, *Ann. de Chim. et de Phys.* 2], **26**, 305.

amide, benzamide, formamide) two isomers; this would be expected from the formula.

$$CH_3\mathit{C}HClCCl_2\mathit{C}H(OH)NHC_2H_3O.$$

But that crotonchloral should be recovered from this in two isomeric forms is inexplicable.

C. TRANSFORMATION OF ISOMERS WITH SEVERAL ASYMMETRIC CARBON ATOMS

As has been stated, compounds containing a single asymmetric carbon, form, on heating, an inactive mixture corresponding to the state of stable equilibrium.

It is otherwise with compounds containing two or more asymmetric atoms. It is evident that here too the inactive mixture corresponds to the state of equilibrium ultimately attained; but this final state is reached in two phases, since in general one of the two asymmetric groups is converted faster than the other. Sometimes, indeed, the conversion of one group may be complete when the other is still unaltered. Beginning, then, with the compound $+A+B$, we shall get first a mixture of $+A+B$ and $+A-B$. It is by no means necessary that the quantities of the two products formed at the end of the first phase should be equal, for the two molecules which are not images of each other are in general of different stability. It is therefore not strange if almost the whole mass becomes converted into $+A-B$, the direction of the rotation being perhaps reversed. In fact, this has been found to take place.

And first let us recall Pasteur's[1] words concerning the transformations in the quinine group:

'Let us consider the three isomers, quinine, quinidine, and quinicine. Quinine is left-handed, quinidine right-handed, both to a considerable degree. Quinicine is right-handed, but, compared with the others, very slightly so. The logical, I had almost said the inevitable, explanation of these results is the following: The quinine molecule is double, and consists of two active bodies, of which one is strongly left-handed, the other very slightly right-handed. This latter is stable on heating, resists transformation into the isomeric group, and, persisting unaltered in quinicine, imparts to this the weak right rotation. The other group, which, on the contrary, is strongly active, becomes inactive when quinine becomes converted by heating into quinicine. Accordingly quinicine would be nothing else than a quinine in which one group has become inactive. Similarly quinicine would be a quinidine in which one group has become inactive; but in quinidine this strongly active group is right-handed instead of left-handed as in quinine, and still combined with that slightly active and stable group which, persisting in quinicine, imparts to this the slight right rotation. I could repeat this word for word with reference to the isomers, cinchonine, cinchonidine, and cinchonicine, which are constituted like the related quinine isomers; for they present exactly the same relations.'

The only difference between these views and those

[1] *Compt. Rend.* 37, 110.

developed above is that in the latter nothing has been said about the so-called groups.

As examples of transformations resulting from change in one of the asymmetric atoms, may be mentioned:

Borneol.—Prepared from ordinary camphor, the product is a mixture in which the left borneol predominates; on heating, almost the whole of this modification is transformed (this is the reason why Montgolfier called it unstable) and produces ordinary right-handed borneol, the sign of the rotation being reversed.

Menthol.—This compound, which contains two asymmetric carbon atoms,

```
        HCC3H7
        /    \
    H2C        CH2
     |          |
    H2C        CO
        \    /
        HCCH3
```

behaves in the same way. Beckmann[1] observed a transformation from left to right rotation on heating to 30° in presence of sulphuric acid.

Gluconic and mannonic acids present, according to Fischer,[2] the same peculiarities, an analogous transformation occurring on heating them with quinoline. Only here in the final condition both isomers are present together, whereas in the former examples one of them almost vanished. Here, too,

[1] *Ann.* **250**, 322. [2] *Ber.* **23**, 800.

we must class camphoric acid, if, with Bamberger,[1] we assume in it two asymmetric carbon atoms. The conversion of right into left acid,[2] which takes place on heating, would then be traceable to the same cause.

We may add, also, the conversion of arabonic into ribonic[3] acid, of galactonic into talonic,[4] of α- into β-gluco-octonic acid,[5] and finally, in all probability, that of left into right ecgonine (p. 62) and of hyoscyamine into atropine (p. 61), both under the influence of alkalies. In the last case the complete disappearance of the activity is remarkable, as in the case of such a slightly active body we might expect transformation in only one of the asymmetric groups; and recently Hesse[6] has announced the activity of atropine, which is especially evident in the sulphate.

It is to be observed further that, so far as is known, the transformation takes place in the part of the molecule richest in oxygen, that is, as near as possible to the carboxyl group when this is present. This is rendered probable in the case of the conversion of left into right ecgonine by the formation of the same anhydrecgonine and of the same tropic acid from both isomers; in the case of the conversion of mannonic into gluconic acid, and of α- and β-gluco-octonic acids by the relations to arabinose and heptose. Moreover, it is well known that in general the presence of oxygen in organic compounds brings

[1] *Ber.* **23**, 218.
[2] Jungfleisch, *Compt. Rend.* **110**, 790.
[3] Fischer, *Ber.* **24**, 4216.
[4] *Ibid.* 2622.
[5] *Ann.* **270**, 64.
[6] *Ibid.* **271**, 100.

about a certain loosening, which often determines the point at which the molecule is attacked and also the breaking up into ions.

It is remarkable that in the transformations considered above, the reverse rotation is in several cases equal to the original rotation:

Stable borneol	$(\alpha) + 37°$ $(\beta) - 37°$
Camphoric acid	$(\alpha) + 46°$ $(\beta) - 46°$
Limonenenitropiperidine	$(\alpha) - 68°$ $(\beta) + 60°$

Mannite and sorbite, both rotating feebly.

Gluconic and mannonic acid, the same.

Arabonic and ribonic acid, the same.

Left menthone, − 28°; right menthone, + 28°.

This equality is not to be confounded with that observed in the case of a single asymmetric carbon atom, and in derivatives it is lacking.

D. SIMPLIFICATION THROUGH SYMMETRY OF THE FORMULA. INACTIVE INDIVISIBLE TYPE

Tartaric acid type.—If we have to do with the presence of asymmetric carbon in a symmetrical formula, the case is simplified. To begin with the simplest case, $\mathit{C}R_1R_2R_3\mathit{C}R_1R_2C_3$, the four symbols given above (p. 57) assume the following form:

No. 1	No. 2	No. 3	No. 4
R_2	R_2	R_2	R_2
R_1 C R_3	R_3 C R_1	R_1 C R_3	R_3 C R_1
R_1 C R_3	R_1 C R_3	R_3 C R_1	R_3 C R_1
R_2	R_2	R_2	R_2

Here, however, No. 1 and No. 4 are identical, as

may be shown with the models, but is evident also from these symbols if we consider that a projection of this kind may be moved round in the plane of the drawing, and therefore may be turned upside down, by turning it through 180° in the direction of the hands of a watch; when this is done No. 1 coincides with No. 4. This configuration is also characterised by the fact that it is symmetrical, as is also shown by the model, but is again expressed in very simple fashion by the symmetry of the projections. There is accordingly no activity to be expected here; it is, then, the 'inactive indivisible type' which results from the symmetry of the formula. The symbols No. 2 and No. 3 are evidently asymmetrical images of each other, and correspond therefore to bodies of opposite activity.

A perfect illustration of this occurs in the isomerism of the tartaric acids. In this group we are, in fact, acquainted with the two isomers of equal and opposite activity, which are represented by the formulæ:

$$\begin{array}{c} CO_2H \\ HOCH \\ HCOH \\ CO_2H \end{array} \quad \text{and} \quad \begin{array}{c} CO_2H \\ HCOH \\ HOCH \\ CO_2H \end{array}$$

as well as the inactive mixture of the two—racemic acid—which was divided by Pasteur. But what especially characterises this case is the existence of an indivisible inactive isomer, which was also discovered by Pasteur, and which some years ago

Przibytek.[1] tried in vain to divide. In fact, such a compound was to be predicted from the formula:

$$\begin{array}{c} CO_2H \\ HCOH \\ HCOH \\ CO_2H \end{array}$$

Erythrite, $CH_2OH(\mathit{C}HOH)_2CH_2OH$, may be cited as a second instance of this inactive indivisible type, since Przibytek[2] has shown that this yields on oxidation the inactive non-racemic tartaric acid. From the constitution of erythrite the possibility of inactivity without divisibility was, in fact, to be expected. Thirdly, we must now add erythrene- or pyrrolylene-bromide, $CH_2Br(\mathit{C}HBr)_2CH_2Br$ (tetrabromobutane), since Griner[3] has converted this into erythrite; the liquid isomeric bromine compound[4] would then represent the racemic mixture.

Here, too, we must mention several compounds whose constitution resembles that of tartaric acid in that they possess a symmetrical formula with two asymmetric carbon atoms. These compounds possess a special interest because they all present a case of isomerism, which, inexplicable according to the old views, is a self-evident necessity of our theory. As in the case of erythrenetetrabromide, these isomers correspond to the inactive indivisible tartaric acid and to racemic acid. Most of these compounds

[1] *Ber.* **17**, 1412. [2] *Ibid.* **20**, 1233.

[3] *Compt. Rend.* **116**, 823.

[4] Henninger, *Compt. Rend.* **104**, 144; Ciamician, *Ber.* **19**, 569; **20**, 3061; **21**, 1430.

have been investigated by Bischoff in his study of the bisubstituted succinic acids possessing the symmetrical formula, $CO_2H(CHX)_2CO_2H$. Such are the dibromo- and isodibromo-succinic acids, dimethyl-,[1] diethyl-,[2] diisopropyl-,[3] and diphenyl-succinic[4] acids, with their derivatives, ethers, anhydrides, &c., which also form isomers. Recent additions to the list are the dimethyldioxyadipic acids,[5] $(CO_2H.CH_3.C.OH.CH_2)_2$, and the thiodilactylic acids[6] $(CO_2H.CH_3.CH)_2S$.

Although up to the present none of these isomers has been divided, yet there is such an intimate connection between their formulæ and those of the tartaric acids that it is difficult to doubt of ultimate success. We have only to substitute methyl, &c., for the hydroxyl of the tartaric acids, and it is more than probable that the isomeric relations of these acids will survive the substitution.

To this class belong also hydro- and isohydrobenzoïn, $C_6H_5(CHOH)_2C_6H_5$, with some derivatives and homologues.[7] These are comparable with tartaric acid, the carboxyl group being now replaced. Finally, we must mention the bromides of nitrostilbene, $NO_2C_6H_4(CHBr)_2C_6H_4NO_2$,[8] and also bi- and isobi-desyl, $C_6H_5(CHCOC_6H_5)_2C_6H_5$.[9]

[1] *Ber.* **18**, 846, 2368; **20**, 2736; **21**, 3170; **22**, 66, 1821.

[2] Bischoff and Hjelt, *Ber.* **20**, 2988, 3078; **21**, 2089; **22**, 67; **23**, 650.

[3] Hell and Mayer, *Ber.* **22**, 56.

[4] Reimer, *Ber.* **14**, 1802; **15**, 2628; Ossipoff, *Compt. Rend.* **109**, 223; Tillmanns, *Ann.* **258**, 87; **259**, 61.

[5] Zelinsky and Isajew, *Ber.* **29**, 819.

[6] Loven, *l.c.* 1132.

[7] Auwers, *Ber.* **24**, 1778.

[8] Bischoff, *Ber.* **21**, 2074.

[9] Knövenagel, *Ber.* **21**, 1359; Garett, **21**, 3107; Fehrlin, **22**, 553.

Glutaric acid type.—What has been said above refers to two asymmetric carbons directly connected. If they are joined by an intermediate atom, we must make a distinction according as this atom is connected with similar or dissimilar groups.

In the first case, for the type $(X)_2C(\mathit{C}R_1R_2R_3)_2$, what has been said above applies equally. We may therefore apply it to the two isomeric dimethyl-[1] and dimethyldioxy-glutaric acids, $H_2C(\mathit{C}HCH_3CO_2H)_2$ and $H_2C(\mathit{C}OHCH_3CO_2H)_2$,[2] to the dimethyladipic acids,[2] $C_2H_4(\mathit{C}HCH_3CO_2H)_2$, as well as to the isomeric bromides of piperylene, $H_2C(\mathit{C}HBrCH_2Br)_2$,[3] and of diallyl, $C_2H_4(\mathit{C}HBrCH_2Br)_2$.[4]

But if there is a difference between the two groups joined to the middle carbon atom, $CXY(CR_1R_2R_3)_2$, then, as Fischer[5] has remarked, a second inactive indivisible modification occurs; this is shown by the difference in the formulæ:

$$\begin{array}{ccc} & R_3 & \\ R_1 & C & R_2 \\ X & C & Y \\ R_1 & C & R_2 \\ & R_3 & \end{array} \quad \text{and} \quad \begin{array}{ccc} & R_3 & \\ R_1 & C & R_2 \\ Y & C & X \\ R_1 & C & R_2 \\ & R_3 & \end{array}$$

This modification has, in fact, been found in the case of the trioxyglutaric acids, $CO_2H(\mathit{C}HOH)_3CO_2H$, and of the corresponding alcohols,

$$CH_2OH(\mathit{C}HOH)_3CH_2OH.$$

[1] Zelinsky, *Ber.* **22**, 2823; Auwers, **23**, 1600; **26**, 4012.

[2] Zelinsky, *Ber.* **22**, 2823; Auwers and V. Meyer, **23**, 295.

[3] Ciamician and Magnanini, *Ber.* **21**, 1434; Wagner, *Ber.* **22**, 3057; *Gazz. Chim.* **16**, 390.

[4] Ciamician and Anderlini, *Ber.* **22**, 2497, 3056.

[5] *Ber.* **24**, 1839.

In the former case we have, besides the active acid ($[\alpha]_D = -23°$) from arabinose, the inactive acid (M.P. 152°) from xylose,[1] and the isomeric unstable inactive acid from ribonic acid, which readily changes into the lactone.

In the case of the alcohols we have side by side the corresponding isomers xylite [1] and adonite.[2]

Also, Zelinsky [3] has prepared three inactive modifications of dimethyltricarballylic acid :

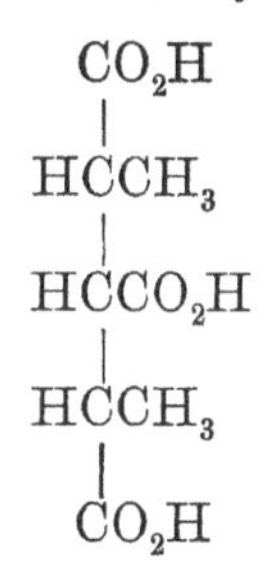

Saccharic acid type.—Finally we come to the symmetrical compounds, which contain four asymmetric carbon atoms, such as saccharic acid. A conspectus of their isomeric relations is afforded by the following symbols, in which the two groupings, HCOH and HOCH, possible with each asymmetric carbon, are indicated by + and −. The sixteen isomers, divided into eight types, are then these :

No. 1	No. 2	No. 3	No. 4
+ −	+ −	+ −	+ −
+ −	+ −	+ −	− +
+ −	+ −	− +	+ −
+ −	− +	+ −	+ −

[1] *Ber.* **26**, 635. [2] *Ibid.* **24**, 538. [3] *Ibid.* **29**, 616.

No. 5	No. 6	No. 7	No. 8
− +	+ −	+ −	+ −
+ −	+ −	− +	− +
+ −	− +	+ −	− +
+ −	− +	− +	+ −

If symmetry exists the two isomers marked No. 1 become identical and inactive; the same with No. 8; the pair No. 2 coincides with No. 5, and No. 3 with No. 4. Hence we have ten isomers, of which two are inactive and indivisible, while the other eight belong to four types. Now, in the case of mannite, $CH_2OH(CHOH)_4CH_2OH$, we have:

Left and right (ordinary) mannite,[1] $[\alpha]_D = +0{\cdot}03$; with boric acid, more strongly right-handed.

Left and right (ordinary) sorbite,[2] slightly active; with borax, $[\alpha]_D = 1{\cdot}4$.

Dulcite, inactive, indivisible.[3]

In the case of the corresponding saccharic acids, indeed, all the six types exist:

Left and right (ordinary) saccharic acid, $[\alpha]_D = 8°$; as lactone, 38°.[4]

Left and right mannosaccharic acid, slightly active; as double lactone, 202°.[5]

Talomucic acid,[6] $[\alpha]_D > +24°$; as lactone, $< +7°$.

Mucic acid, inactive, indivisible.[7]

Allomucic acid, inactive, indivisible.[8]

[1] Kiliani, *Ber.* **20**, 2714.
[2] Fischer and Stahel, *Ber.* **26**, 2144.
[3] *Ber.* **25**, 2564, 1247.
[4] Tollens's *Kohlehydrate.*
[5] *Ber.* **24**, 541, 3628.
[6] Fischer, *Ber.* **24**, 3622.
[7] Fischer, *l.c.* **25**, 1247.
[8] Fischer, *l.c.* **24**, 2136

CHAPTER IV

DETERMINATION OF THE POSITION OF THE RADICALS IN STEREOMERS

WHEN the number of the isomers actually existing (which, in the cases we have been considering, may be called stereomers) agrees with the theory, we are confronted with a problem like that which we have to solve in the aromatic series, when we assign to each of three derivatives one of the three symbols 1, 2, 1, 3, 1, 4. At present this problem can be solved only partially: which of the two enantiomorphous formulæ corresponds to, say, the left-rotating compound, is undecided. When, however, there are several carbon atoms the case is different. We have already mentioned such types. In the case of tartaric acid, *e.g.* (p. 75), the symbol

$$\begin{array}{c} CO_2H \\ HCOH \\ HCOH \\ CO_2H \end{array}$$

was chosen on account of its symmetry as the expression for the 'inactive indivisible type,' while the two other formulæ remained for the right- and left-acids; to decide between these last is, however, impossible. It is especially in the sugar group that the determination of configuration, in this sense, has been carried out by Fischer.[1] In now discussing the special data and results, since we can choose the formula for the type only, and not for the right- or left-handed product in question, we find that the number of symbols to be distributed is reduced by half, which greatly simplifies the discussion. In what follows, therefore, the mirror-images, such as

CO_2H	CO_2H
HCOH	HOCH
HOCH	HCOH
CO_2H	CO_2H

represent the same tartaric acid—in this case the active one.

In order now to facilitate the review of the sugar-derivatives we will take in succession first the simplest, the tetroses, $COH(CHOH)_2CH_2OH$, then the pentoses, $COH(CHOH)_3CH_2OH$, and finally the glucoses, $COH(CH.OH)_4CH_2OH$. Then we have 4, 8, and 16 isomers, or 2, 4, and 8 types, and the first two are directly connected with the tartaric acids, their symbols being

H_2COH		H_2COH
HCOH	and	HCOH
HCOH		HOCH
COH		COH

and the substances represented by the first symbol giving inactive, indivisible tartaric acid, those represented by the other giving the left- or right-

[1] *Ber.* **24**, 1836, 2684.

acid. This tetrose has been recently obtained from arabinose.[1]

The following table gives, so to speak, the development of the pentoses and glucoses from these two tetroses, according to the experimental results of Kiliani and Fischer.

Tetroses ($C_4H_8O_4$)

I	II
H_2COH	H_2COH
HCOH	HCOH
HCOH	HOCH
OCH	OCH
(corresponds to inactive tartaric acid)	(corresponds to active tartaric acid)

Pentoses ($C_5H_{10}O_5$)

I		II	
A	B	A	B
H_2COH	H_2COH	H_2COH	H_2COH
HCOH	HCOH	HCOH	HCOH
HCOH	HCOH	HOCH	HOCH
HCOH	HOCH	HOCH	HCOH
OCH	OCH	OCH	OCH
Ribose	Arabinose	Lyxose	Xylose

Glucoses ($C_6H_{12}O_6$)

I A		I B	
α	β	α	β
H_2COH	H_2COH	H_2COH	H_2COH
HCOH	HCOH	HCOH	HCOH
HCOH	HCOH	HCOH	HCOH
HCOH	HCOH	HOCH	HOCH
HCOH	HOCH	HCOH	HOCH
OCH	OCH	OCH	OCH
Allomucic acid	Talomucic acid Talose ?	Saccharic acid Glucose Sorbite	Manno-saccharic Mannose Mannite

[1] Fischer, *Ber.* **26**, 740.

Glucoses ($C_6H_{12}O_6$)—*continued.*

II A		II B	
α	β	α	β
H_2COH	H_2COH	H_2COH	H_2COH
HCOH	HCOH	HCOH	HCOH
HOCH	HOCH	HOCH	HOCH
HOCH	HOCH	HCOH	HCOH
HCOH	HOCH	HCOH	HOCH
OCH	OCH	OCH	OCH
Mucic acid	Talomucic acid	Saccharic acid	Idosaccharic acid
Galactose	Talose	Gulose	Idose
Dulcite		Sorbite	Idite

The respective tetrose or pentose is enriched by CHOH, by addition of hydrogen cyanide, conversion of cyanogen into carboxyl, and, finally, reduction of the resulting acid (or rather of its lactone), the group OCH being converted successively into HO*C*HCN, HO*C*HCO$_2$H, and HO*C*HCOH. The formation of two isomers is then indicated by the symbols; they are distinguished in the case of the pentoses by A and B, in the case of the glucoses by *α* and *β*.

It is noteworthy that, thanks to the recent researches of Wohl,[1] the process has been carried out in the opposite direction, the oxime HOCHNOH being formed, and hydrogen cyanide removed from this by ammoniacal silver oxide.

We have now to find for each of the known isomers its place in the table.

Pentose group.—In the first place we have to refer the four pentose types to the symmetrical acids. Of the three possible types,

[1] *Ber.* **26**, 740.

$$\begin{array}{ccc}
CO_2H & CO_2H & CO_2H \\
HCOH & HCOH & HCOH \\
HCOH & HCOH & HOCH \\
HCOH & HOCH & HCOH \\
CO_2H & CO_2H & CO_2H
\end{array}$$

only the second is active, accordingly, for arabinose, which on oxidation yields this acid, the choice lies between :

$$\begin{array}{ccc}
CH_2OH & & COH \\
HCOH & & HCOH \\
HCOH & \text{and} & HCOH \\
HOCH & & HOCH \\
COH & & CH_2OH
\end{array}$$

i.e. between I B and II A in the table.

Now, in the Kiliani-Fischer reaction arabinose yields glucose and mannose,[1] which therefore are represented either by I B α, β, or by II A α, β. By oxidising these to the symmetrical acids,

$$CO_2H(CHOH)_4CO_2H,$$

we get saccharic and mannosaccharic acids[2] respectively; both are active, which agrees only with I B α, β, since II A α would give an inactive isomer. Accordingly the arabinose formula must be I B :

$$\begin{array}{c}
H_2COH \\
HCOH \\
HCOH \\
HOCH \\
OCH
\end{array}$$

[1] *Ber.* **23**, 799. [2] *Ibid.* **24**, 539.

while to the recently discovered lyxose [1] which yields mucic acid (II A *a*) on oxidation we assign the formula II A:

$$\begin{array}{c} H_2COH \\ HCOH \\ HOCH \\ HOCH \\ OCH \end{array}$$

Now, from arabonic acid, $H_2COH(HCOH)_3CO_2H$, which corresponds to arabinose, we get, on heating, ribonic acid,[2] and we must assume that the transformation takes place in the neighbourhood of the highly oxygenated carboxyl-group. Ribonic acid is, then:

$$\begin{array}{c} H_2COH \\ HCOH \\ HCOH \\ HCOH \\ CO_2H \end{array}$$

Further, the configuration of adonite,[3]

$$CH_2OH(CHOH)_3CH_2OH,$$

the reduction-product of ribose, must be taken as corresponding with the above; while for xylite [4] and xylose the last possibility remains:

$$\begin{array}{c} H_2COH \\ HCOH \\ HOCH \\ HCOH \\ OCH \end{array}$$

[1] *Ber.* **29**, 581. [2] *Ibid.* 4214. [3] *Ibid.* **26**, 636. [4] *Ibid.* **24**, 528.

As for lyxonic acid, which corresponds to lyxose, it is obtained from xylonic acid just as ribonic from arabonic acid, by a transformation in the neighbourhood of the carboxyl group. The formula above given for lyxose is thus confirmed.

In the group of the pentoses, of the corresponding pentatomic alcohols, alcohol acids, and trioxyglutaric acids, all the configurations are, then, determined:

Ribose. Ribonic acid. Adonite (inact.). Inact. trioxygl.	Arabinose. Arabonic acid. Arabite (act.). Act. trioxygl.
H_2COH	H_2COH
HCOH	HCOH
HCOH	HCOH
HCOH	HOCH
OCH	OCH

Lyxose. Lyxonic acid. Act. trioxygl.	Xylose. Xylonic acid. Xylite (inact.). Inact. isomeric trioxygl.
H_2COH	H_2COH
HCOH	HCOH
HOCOH	HOCH
HOCOH	HCOH
OCH	OCH

Glucose group.—It has already been mentioned that glucose and mannose have the formulæ I B *a*, *β*. The choice is rendered possible by the fact that the same saccharic acid which results from the oxidation of glucose is obtained also from an isomeric gulose;[1] only the formula I B *a* admits of such an isomer, and therefore the configuration of mannose and gulose, of mannosaccharic and saccharic acid, is at once settled, as well as that of the corresponding mannite and sorbite, which are formed on reducing mannose[2] and glucose[3] respectively.

Glucose. Saccharic acid. Sorbite.	Mannose. Manno-saccharic acid. Mannite.	Gulose. Saccharic acid. Sorbite.
H_2COH	H_2COH	H_2COH
HCOH	HCOH	HCOH
HCOH	HCOH	HOCH
HOCH	HOCH	HCOH
HCOH	HOCH	HCOH
OCH	OCH	OCH

At the same time this determines the configuration of levulose. The constitution is, according to Kiliani, $H_2COH(HCOH)_3COCH_2OH$. Now, as this yields on reduction sorbite and mannite[4]:

H_2COH	H_2COH	H_2COH
HCOH	HCOH	HCOH
HCOH	HCOH	HCOH
HOCH	HOCH	HOCH
HCOH	HOCH	CO
H_2COH	H_2COH	H_2COH
Sorbite.	Mannite.	Levulose.

it must possess the third formula.

Glucose group, mucic acid derivatives.—Since it is

[1] Fischer, *Ber.* **24**, 521. [2] *Ber.* **22**, 365; **24**, 539.
[3] Meunier, Delachanal, *Compt. Rend.* **111**, 49, 51.
[4] *Ber.* **23**, 2611.

proved that mucic acid, $CO_2H(HCOH)_4CO_2H$, and the corresponding dulcite belong to the 'inactive indivisible type,'[1] we have only to choose between the two following configurations for the acid:

CO_2H		CO_2H
HCOH		HCOH
HCOH	and	HOCH
HCOH		HOCH
HCOH		HCOH
CO_2H		CO_2H

Then we have for galactonic acid,

$$CH_2OH(CHOH)CO_2H$$

(and galactose), two possibilities. Now, this acid is converted into talonic acid (and talose) by heating the quinoline- and pyridine-salt,[2] and the transformation must be supposed to take place in the HCOH group next to the carboxyl. Talonic acid is accordingly:

H_2COH		H_2COH
HCOH		HCOH
HCOH	or	HOCH
HCOH		HOCH
HOCH		HOCH
CO_2H		CO_2H

But this determines the configuration of the talomucic acid obtained by oxidation. Of the four active types we have now determined three, saccharic acid by the configuration of glucose, mannosaccharic

[1] *Ber.* 25, 1247. [2] *Ibid.* 24, 1841.

acid by that of mannose, and also talomucic acid. We have then :

CO_2H	CO_2H	CO_2H	CO_2H	CO_2H	CO_2H
HCOH	HCOH	HCOH	HCOH	HCOH	HCOH
HCOH	HCOH	HCOH	HCOH	HOCH	HOCH
HCOH	HCOH	HOCH	HOCH	HOCH	HCOH
HCOH	HOCH	HCOH	HOCH	HCOH	HOCH
CO_2H	CO_2H	CO_2H	CO_2H	CO_2H	CO_2H
Inactive allo-mucic acid.	Talo-mucic (?) acid.	Saccharic acid (Sorbite).	Manno-saccharic acid (Mannite).	Inactive mucic acid (Dulcite).	Ido-saccharic acid (Idite).

Finally, the following table is appended to afford a conspectus of the relations thus established. It contains, of course, only half of the possible isomers; the other half corresponds to the mirror-images. Fischer has proposed to distinguish by the letters *d*- and *l*- the two groups which belong together, and has chosen *d*- for that which contains the long-known dextroglucose; the rotations given in the table are based on this plan.

It must be added that some compounds have been included of which only the enantiomorphous form is known, *e.g.* xylose, arabonic and ribonic acid; in such cases, however, we need not scruple to reverse the sign of the rotation. The formulæ thus obtained have done excellent service as guides in following out the relations of these compounds. They explain, *e.g.*:

1. That levulose is broken up on oxidation into glycollic acid and *inactive* tartaric acid.[1]

[1] Kiliani, *Ber.* **14**, 2530.

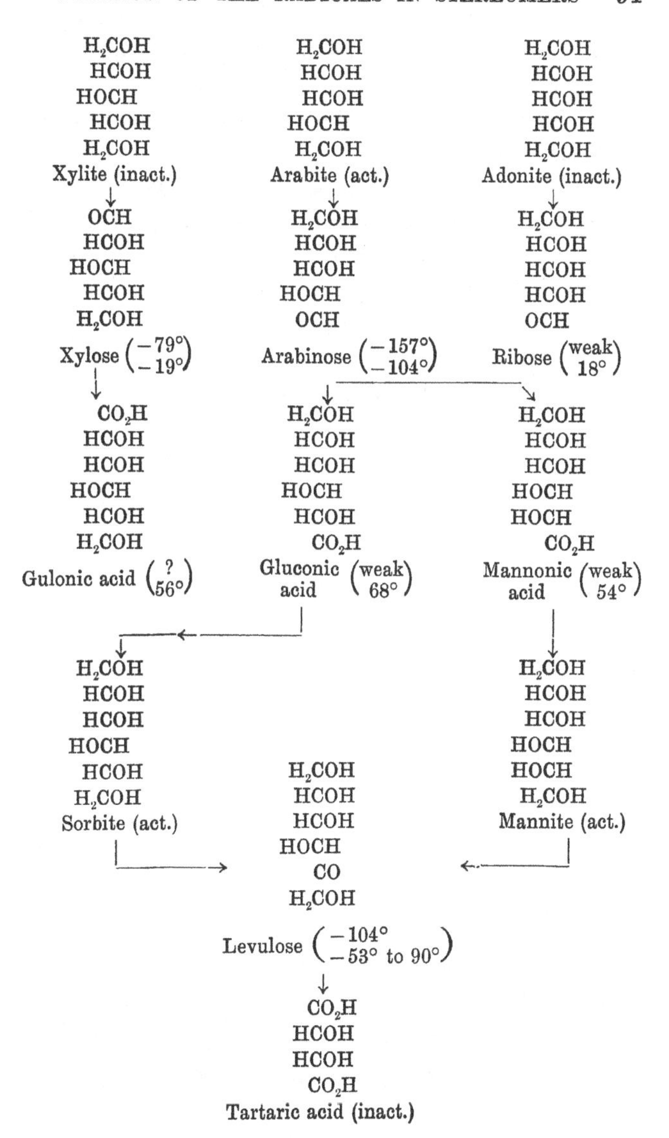
H2COH
HCOH
HOCH
HCOH
H2COH
Xylite (inact.)
H2COH
HCOH
HCOH
HOCH
H2COH
Arabite (act.)
H2COH
HCOH
HCOH
HCOH
H2COH
Adonite (inact.)
OCH
HCOH
HOCH
HCOH
H2COH
Xylose (−79° −19°)
H2COH
HCOH
HCOH
HOCH
OCH
Arabinose (−157° −104°)
H2COH
HCOH
HCOH
HCOH
OCH
Ribose (weak 18°)
CO2H
HCOH
HCOH
HOCH
HCOH
H2COH
Gulonic acid (? 56°)
H2COH
HCOH
HCOH
HOCH
HCOH
CO2H
Gluconic acid (weak 68°)
H2COH
HCOH
HCOH
HOCH
HOCH
CO2H
Mannonic acid (weak 54°)
H2COH
HCOH
HCOH
HOCH
HCOH
H2COH
Sorbite (act.)
H2COH
HCOH
HCOH
HOCH
HOCH
H2COH
Mannite (act.)
H2COH
HCOH
HCOH
HOCH
CO
H2COH
Levulose (−104° −53° to 90°)
CO2H
HCOH
HCOH
CO2H
Tartaric acid (inact.)

2. That glucose, by means of its osazone, can be changed into levulose.[1]

3. That mannite forms on oxidation mannose and levulose.[2]

4. That glucose, on being treated according to the method of Kiliani-Fischer, gives a glucoheptonic acid in two isomers, of which one forms on oxidation an inactive, indivisible pentoxypimelic acid,[3] and so on.

[1] *Ber.* 22, 94.
[2] Dafert, *ibid.* 17, 227.
[3] Fischer, *Ann.* 270, 64.

CHAPTER V

THE UNSATURATED CARBON COMPOUNDS

I. Statement of the Fundamental Idea

Historical.—In planning this chapter for the new edition, it was of special importance to make plain the present position of the theory.

Having hitherto considered chiefly the derivatives of methane, CH_4, we have now to do with those of ethylene, C_2H_4. The problem is here more complicated, since there are now six atoms whose relative position is to be considered, whereas before there were only five; and accordingly we find the position of affairs less satisfactory.

With regard to the asymmetric carbon atom, Le Bel's conceptions and mine led to the same result. There was here at least the possibility of a difference. My fundamental idea was the tetrahedral grouping, that is to say, any force—cause so far unknown—proceeding from the carbon atom and tending to drive the groups united with carbon as far away from one another as possible, that is, to bring them into the tetrahedral position. Although it did not necessarily follow that the tetrahedron must be regular because the mutual action of the different groups might vary its form somewhat, yet the tendency to form the regular tetrahedron remained, and in the case of identity among the groups, as in CH_4, the tendency was realised.

To Le Bel, the asymmetry of the tetrahedron with different, and the symmetry with identical groups, seemed established, CH_4, *e.g.* might be a regular four-sided pyramid, with carbon at the summit and the hydrogens at the corners of the square base.[1]

At present this cannot be decided. So that as regards methane derivatives we are practically agreed.

With substituted ethylenes the case is different. I had at once concluded, as will presently be set forth in detail, that the four groups are in one plane, in which lie the carbon atoms also; here, then, there is never any possibility of dissymmetry but only of another kind of isomerism, like that of fumaric and maleïc acid. To Le Bel the question seemed an open one; experiment would have to decide. It was only after some time[2] that, influenced by the researches of Kekulé and Auschütz, he declared himself in favour of my view.

But later another change occurred. Doubts arose in Le Bel's mind on account of indications of asymmetry, *i.e.* optical activity among substituted ethylenes. He had observed[3] that a solution of citraconic acid, $CH_3C(CO_2H){=}CH(CO_2H)$, acquires

[1] *Bull. Soc. Chim.* [3] **3**, 788; *Compt. Rend.* **114**, 304.
[2] *Bull. Soc. Chim.* **37**, 300. [3] *Ibid.* [3] **7**, 164.

activity through the growth of fungi. If active citraconic acid had been thus formed, the activity of ethylene derivatives was proved; it was found,[1] however, that the activity was due to the formation, by addition of water, of methylmalic acid

$$\begin{array}{l} \mathit{C}H.\ CH_3.\ CO_2H \\ | \\ \mathit{C}H.\ OH.\ CO_2H, \end{array}$$

and this no doubt accounts for the active product formed in the case of mesaconic acid also; allyl alcohol and *α*-crotonic acid gave no active product; the results in the case of fumaric and maleïc acid were doubtful.

There could be adduced then only the supposed activity of styrolene, $C_6H_5.HC{=}CH_2$, and of chlorofumaric and chloromaleïc acid, $CO_2H.ClC{=}CH.CO_2H$ (Perkin).[2] My researches (p. 20) had, however, already rendered the activity of styrol very doubtful, and presently Walden's[3] investigation showed the observation of Perkin to be positively incorrect. It remains only to state the facts which make the activity of ethylene derivatives seem to me improbable.

In the first place there are numerous ethylene derivatives occurring in nature, among them such as have two different groups attached to each of the two carbon atoms; tiglic acid,

$$CH_3CH{=}C(CH_3)CO_2H,$$

and numerous compounds of the oleic series, fumaric

[1] Le Bel, *Bull. Soc. Chim.* [3] **11**, 292.

[2] *J. Chem. Soc. Trans.* 1888, 695.

[3] *Ber.* **26**, 508.

acid, cinnamic acid, coumaric acid, anethole, asarone, piperine. They are all inactive.

In the second place I may mention the statements published long since as to the formation of ethylene derivatives from active compounds; the activity uniformly disappears:

Inactive fumaric and maleïc acids from active malic acid;

Inactive chloro-fumaric and maleïc acids from active tartaric acid;[1]

Inactive crotonic acid from active β-oxybutyric acid;[2]

Inactive furfurol from active arabinose and xylose;[1]

Inactive coniferyl alcohol from active coniferine.

In the third place, fumaric acid could not be divided by Auschütz and Hintze,[3] while, according to a private communication from Walden, the growth of microbes in maleïc acid gave a similar negative result.

Finally, at my request, Liebermann has converted his active cinnamic acid dibromide,

$$C_6H_5(CHBr)_2CO_2H,$$

at a low temperature, into bromo-cinnamic acid, $C_6H_5CBrCHCO_2H$, and Walden his active chloro-succinic acid into fumaric acid. Both derivatives proved inactive. At present, then, no reason for a change of opinion is apparent.

[1] van 't Hoff, *Ber.* **10**, 1620. Walden, *l.c.*
[2] Deichmüller, Szymanski, Tollens, *Ann.* **228**, 95.
[3] *Ibid.* **239**, 164.

Relative position of the groups attached to doubly linked carbon; cessation of free rotation.—The fundamental idea that the four groups connected with carbon occupy the corners of a tetrahedron, requires, in order that it may be applied to doubly linked

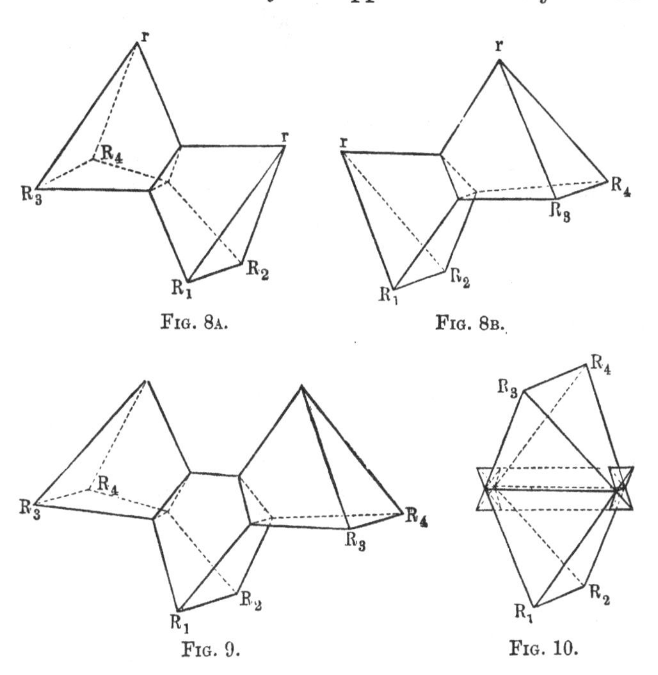

FIG. 8A. FIG. 8B.

FIG. 9. FIG. 10.

carbon, a clear conception of the nature of this linkage. As to this, we assume that the relative position of the two connected tetrahedra corresponds with that which we assumed in the case of the single bond; but now two corners of the tetrahedron play the part which formerly was reserved for one

Further, having regard to the now universally assumed equality of the carbon affinities, each of the two tetrahedron corners must play in the act of combination a perfectly identical part. In order to arrive at the grouping corresponding to this view, we must find that relative position of the tetrahedra which lies half-way between the two cases of single linking in which the one or the other pair of corners is joined. Let us consider, then, a compound, $CR_1R_2rCR_3R_4r$, and represent it in the two different forms which are obtained if we leave the group CR_1R_2 in the same position, but attach to it the groups r and CR_3R_4r in two different ways, as shown in figs. 8A and 8B.

Passing now to the unsaturated compound $CR_1R_2=CR_3R_4$, we have to eliminate the two r groups and to place CR_3R_4 in a position half-way between the two cases. This position is easily perceived if we unite the two cases in a single figure (9). In fact we arrive at the intermediate position shown in fig. 10, in which the groups R_3 and R_4, and R_1, R_2 are in one plane, with regard to which the two positions shown in fig. 9 are symmetrical.

Graphic representation.—The grouping thus arrived at can be represented with the utmost simplicity by using the following formula:

$$\begin{array}{c} R_1CR_2 \\ \| \\ R_3CR_4 \end{array}$$

Prediction of cases of isomerism.—Besides the

above-described relative position of the four groups R_1, R_2, R_3, R_4, there is another, which also satisfies the conditions laid down, but yet is not identical with the first. The groups R_1 and R_2 may lie in one plane with R_3 and R_4, each joined to the same carbon atom as before, but with the difference that R_1 is opposite R_4, and R_2 opposite R_3 :

$$\begin{array}{c} R_1CR_2 \\ \| \\ R_4CR_3 \end{array}$$

Consequently there must be here an isomerism unforeseen by the old formulæ, and it is clear that this isomerism must be expected in every case where the groups attached to the same carbon, R_1, R_2 and R_3, R_4, are different, and this whether the groups attached to different carbons are alike or not, so that *e.g.* the same isomerism would occur in the case of $CR_1R_2CR_1R_2$.

II. Confirmation of the Fundamental Idea

General character of the isomerism to be expected in the case of doubly linked carbon.—In the first place we must call attention to the nature of this isomerism, because a marked difference is to be expected between this and the isomerism due to the presence of asymmetric carbon. For, according to the views just set forth, there is here neither dissymmetry nor enantiomorphism in structure, so that we should not expect either the rotatory power, in opposite directions in the two cases, nor the peculiar hemihedral crystalline form which accompanies this optical behaviour; and, as we shall see, these two properties are altogether lacking. But we must expect to find a profound difference in the other properties of the two isomers. Whereas there was in this respect complete identity between the two isomers of opposite activity, an identity harmonising perfectly with the assumed equality of their molecular dimensions, this identity must for the very same reasons be lacking in the present case, because on the one hand we must assume a difference in the physical properties in general (difference in the quantities *a* and *b* of van der Waals' theory), in specific gravity, melting- and boiling-point, solubility, &c., while, on the other hand, a chemical difference is to be expected, that is to say a difference in stability, heat of formation, &c.[1]

We may classify the cases coming within this category as follows:

A. SIMPLE ETHYLENE DERIVATIVES

Monochloropropylene[2] . . .	$CH_3CHCl = CH_2$
Bromopseudobutylene[3] . . .	$CH_3CBr = CHCH_3$
Crotonylenebromide[3] . . .	$CH_3CH_2CBr = CHBr$
" "[4] . . .	$CH_3CBr = CBrCH_3$
Tolanechloride[5]	$C_6H_5CCl = CClC_6H_5$
Tolanebromide[5]	$C_6H_5CBr = CBrC_6H_5$

[1] A marked physiological difference has been observed by Fodera (*Ref. Chem. Ztg.* 19, Repertorium 407). Injection of maleïc acid kills a dog quickly, whereas the like quantity of fumaric acid has no poisonous action. As to differences in refractive and dispersive power, *see* Brühl, *Ber.* 29, 2902.

[2] Wislicenus, *Ber.* 20, 1008.

[3] Hölz, *Ann.* 250, 230.

[4] Faworsky, *Journ. f. prakt. Chem.* 1890, 149.

[5] Zinin, *Ber.* 4, 288; Limpricht, *ibid.* 379; Blank, *Ann.* 248, 20; Eiloart, *Am. Chem. J.* 12, 231.

o-Dinitrostilbene[1]	$C_6H_4NO_2CH = CHC_6H_4NO_2$
Apiol and Isapiol[2]	$C_9H_9O_4CH = CHCH_3$
Anethol[3]	$C_6H_4OCH_3CH = CHCH_3$
Nitrostyrol[4]	$C_6H_5CH = CHNO_2$

B. UNSATURATED MONOBASIC ACIDS (ACRYLIC ACID SERIES)

β-bromacrylic acid[5]	$CHBr = CHCO_2H$
β-iod[6] ,, ,,	$CHI = CHCO_2H$
Furfuracrylic acid	$CHC_4H_3O = CHCO_2H$
Crotonic and isocrotonic acid	$CH_3CH = CHCO_2H$
β-chloro- ,, ,, ,, [7]	$CH_3CCl = CHCO_2H$
α-chloro- ,, ,, ,, [8]	$CH_3CH = CClCO_2H$
α- and β-brom-acid[7]	$CH_3CH = CBrCO_2H$
β-thioethyl, thiophenyl, and thiobenzyl acid[9]	$CH_3C(SC_2H_5) = CHCO_2H$
Bromomethacrylic acid[10]	$CHBr = C(CH_3)CO_2H$
Tiglic and angelic acid[11]	$CH_3CH = C(CH_3)CO_2H$
Hydrosorbic acid[12]	$C_3H_7CH = CHCO_2H$
Hypogæic and gaidic acid	$CH_3CH = CH(C_{13}H_{25}O_2)$
Oleic and elaidic acid	$CH_3CH = CH(C_{15}H_{29}O_2)$
Erucic and brassic acid[13]	$CH_3CH = CH(C_{19}H_{37}O_2)$

C. AROMATIC MONOBASIC ACIDS (CINNAMIC ACID SERIES)

Cinnamic and isocinnamic acid[14]	$C_6H_5CH = CHCO_2H$
α-bromocinnamic acid[7 15]	$C_6H_5CH = CBrCO_2H$
β- ,, ,, ,, [7 15]	$C_6H_5CBr = CHCO_2H$
Dibromocinnamic acid[16]	$C_6H_5CBr = CBrCO_2H$

---

[1] Bischoff, *Ber.* **21**, 2073; Thiele and Dimroth, *ibid.* **28**, 1411.
[2] Ciamician, *ibid.* 1621. [3] Beilstein. [4] *Ber.* **19**, 1936.
[5] Michael, *ibid.* 1385. [6] Stolz, *ibid.* 542.
[7] Mirbach, *ibid.* 1384; Authenrieth, *ibid.* **29**, 1645, 1670.
[8] Wislicenus, *ibid.* **20**, 1008.
[9] Authenrieth, *ibid.* 1531; **29**, 1639.
[10] Fittig, *Ann.* **206**, 16. [11] *Ibid.* **216**, 16.
[12] *Ibid.* **200**, 51; *Ber.* **15**, 618.
[13] Holt, *ibid.* **24**, 4126. [14] Liebermann, *ibid.* **23**, 141.
[15] Erlenmeyer, *ibid.* **19**, 1936. [16] Roser, *ibid.* **20**, 1576.

α-chlorocinnamic acid [1]	$C_6H_5CH = CClCO_2H$
o-, m- and p-nitrophenylcinnamic acid	$NO_2C_6H_4CH = C(C_6H_5)CO_2H$
Cumaric acid [2]	$C_6H_4(OH)CH = CHCO_2H$
Methyl- and ethyl-cumaric acid [3] .	$C_6H_4(OMe)CH = CHCO_2H$
α- and β-hydropiperic acid [4] . .	$(C_7H_5O_2)CH = CH(C_2H_4CO_2H)$
benzallævulic acid [5] . . .	$C_6H_5CH = C(COCH_3)CH_2CO_2H$

D. DIBASIC ACIDS (FUMARIC ACID SERIES)

Fumaric and maleïc acid . . .	$CO_2HCH = CHCO_2H$
Halogen derivatives	$CO_2HCX = CYCO_2H$
Hydroxyl derivatives [6] . . .	$CO_2HC(OH) = C(OH)CO_2H$
Citra- and mesaconic acid . .	$CH_3CCO_2H = CHCO_2H$
Dimethylfumaric [7] and maleïc [8] acid	$CH_3.CCO_2H = C.CH_3.CO_2H$
Diphenylfumaric and maleïc acid [9] .	$CO_2HC(C_6H_5) = C(C_6H_5)CO_2H$
Camphoric acid (p. 60).	

Perhaps in some of the above cases it is not established to the satisfaction of everyone that both isomers possess the same constitution ; and in a few cases the existence of the isomerism is questioned. On the other hand some isomers have probably been overlooked, and all chemists, even those who are opposed to stereochemical conceptions, are convinced that with doubly linked carbon, when the attached groups are different, isomerism results.

Camphoric acid (see p. 60) is included in the list because it is possible that the four known isomers are due to a combination of asymmetry with double linkage.

[1] Plöchl, *Ber.* **15**, 1946. [2] Roser, *ibid.* 2348.
[3] Fittig, *Ann.* **206**, 16. [4] *Ibid.* **216**, 171.
[5] Erdmann, *ibid.* **258**, 130.
[6] Fenton, *J. Chem. Soc.* 1896, 546. [7] Fittig, *Ber.* **29**, 1842.
[8] Exists only as the anhydride, pyrocinchonic acid.
[9] Rügheimer, *Ber.* **15**, 1625.

ALLYLENE TYPE—SECOND CASE OF OPTICAL ACTIVITY

The following prediction may here be repeated *verbatim* from the earlier edition.

The combination $(R_1R_2)C = C = C(R_3R_4)$ is represented in fig. 11. Here, too, we shall have two isomers, as follows from the difference between

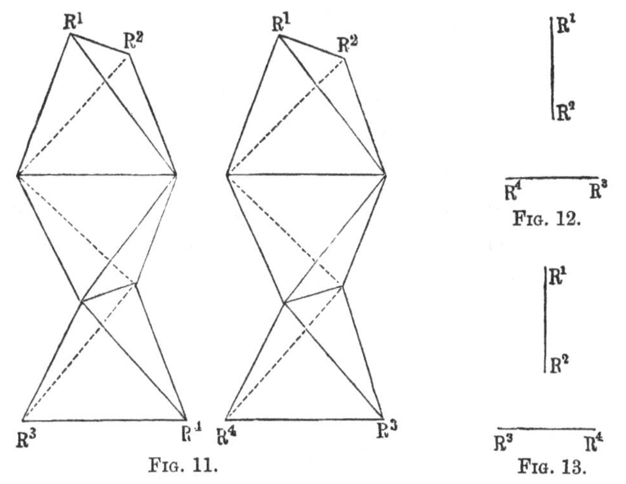

Fig. 11.

Fig. 12.

Fig. 13.

figs. 12 and 13, figures which result from the application of the graphic method above mentioned. The conditions with regard to the equality or difference of the attached groups are the same as in the preceding case. The models of the isomers are in this case enantiomorphous.

It is evident that the case of

$$(R_1R_2)C = C = C = C(R_3R_4),$$

or, in general,

$$(R_1R_2)C = C_{2n} = C(R_3R_4),$$

is the same as the case of

$$(R_1R_2)C = C(R_3R_4).$$

Of combinations of this kind there always exist two isomers when there is a difference between the groups R_1 and R_2, as well as between R_3 and R_4. The models of the isomers are not enantiomorphous.

On the other hand, the case of

$$(R_1R_2)C = C = C = C = C(R_3R_4),$$

or, in general,

$$(R_1R_2)C = C_{2n+1} = C(R_3R_4),$$

is the same as the case of

$$(R_1R_2)C = C = C(R_3R_4).$$

Thus, of these combinations also, there are always two isomers when there is a difference between R_1 and R_2 as well as between R_3 and R_4. The models of the isomers are enantiomorphous.

Treble linkage.—Two carbon atoms, trebly linked, which, according to the ordinary formulæ, are expressed by the symbol $C \equiv C$, may, on the hypothesis of the equality of the bonds, be represented by two tetrahedra having three corners in common, and therefore having a surface of each coinciding, so that they form a double three-sided pyramid (fig. 14). R_1 and R_2 are the monad groups by which the two free affinities of the system are saturated. In this case a difference in the relative position of the saturating

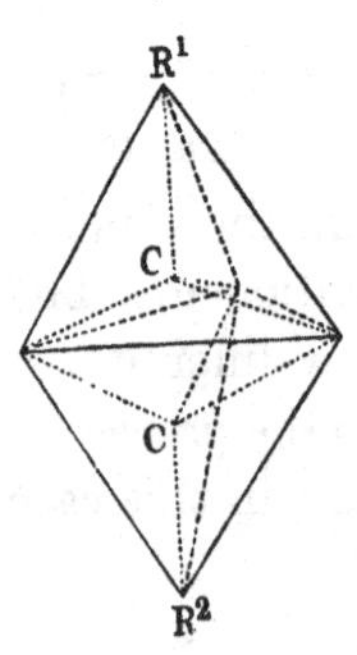

FIG. 14.

groups is not possible, and the possibility of isomerism is, in accordance with the prevalent views, excluded.

III. Determination of Relative Position in Unsaturated Compounds

Whereas, for the isomers of opposite rotatory power, it is at present impossible to decide which structure a given modification possesses, the state of things is much more favourable in this respect for isomers which have a double bond, like fumaric and maleïc acid. For these substances the question was settled at the outset, and it is due principally to the development of the subject by J. Wislicenus[1] that considerations of this kind have met with general recognition. We have to do in particular with two principles which seem capable of solving this problem, viz. with the mechanism of addition which forms and transforms the isomers, and with the mutual influence of the groups within the molecule.

As regards the *mechanism of addition* we can use the same principle which governs every determination of chemical structure by the aid of the formation and transformation of known compounds, and which consists in the assumption that in chemical processes the atomic structure remains as far as possible unaltered.

According to this principle, then, it must be expected that on making an addition to bodies with a triple carbon linkage, two of the three connected

[1] *Abhandl. der Königl. Sächs. Ges.* 1887.

pairs of corners will remain unaltered.[1] Hence follows that acetylene dicarboxylic acid,

$$C(CO_2H)C(CO_2H),$$

e.g. on addition of bromine, will yield the compound

$$\begin{array}{c} CO_2HCBr \\ \| \\ CO_2HCBr \end{array}$$

In opposition to this Bandrowski[2] had proved the formation of dibromofumaric acid. But Wislicenus, guided by these theoretical views, repeated the experiments and showed that, in fact, dibromomaleïc acid is formed.

Later these experiments were taken up by Michael;[3] in a detailed paper, where the theory in question is critically discussed, the formation of dibromomaleïc acid up to 28 or 33 per cent. is confirmed, but there was found also about double the quantity of the isomeric substance. The objection arising from the formation of this latter is, however, as Wislicenus also observes, not important; it is well known, in fact, with what ease maleïc acid changes into fumaric;[4] under the influence of light and a trace of bromine, I have myself seen this transformation ensue so rapidly that it was possible to take a photograph in fumaric acid from the solution of maleïc acid; moreover, in Michael's experiments the *status nascens* has to be taken into account.

[1] van 't Hoff, *Études de dyn. chim.* 1884, 100.
[2] *Ber.* **12**, 2122.
[3] *J. prakt. Chem.* **46**, 210.
[4] Compare Wislicenus, *Ber.* **29**, Ref. 1080.

In the case of addition to substances with a double linkage, the principle above-mentioned demands that, of the two connected pairs of corners, one shall remain unaltered. If, then, we add hydroxyl to fumaric and maleïc acid we obtain[1] from

$$\begin{array}{c} CO_2HCH \\ \| \\ HCCO_2H \end{array} \quad \text{or} \quad \begin{array}{c} HCCO_2H \\ \| \\ CO_2HCH \end{array}$$

and

$$\begin{array}{c} CO_2HCH \\ \| \\ CO_2HCH \end{array} \quad \text{or} \quad \begin{array}{c} HCCO_2H \\ \| \\ HCCO_2H \end{array}$$

the compounds

$$\begin{array}{c} OH \\ CO_2HCH \\ HCCO_2H \\ OH \end{array} \quad \text{or} \quad \begin{array}{c} OH \\ HCCO_2H \\ CO_2HCH \\ OH \end{array}$$

and

$$\begin{array}{c} OH \\ CO_2HCH \\ CO_2HCH \\ OH \end{array} \quad \text{or} \quad \begin{array}{c} OH \\ HCCO_2H \\ HCCO_2H \\ OH \end{array}$$

—that is to say, we get racemic acid in the first case, and inactive tartaric acid in the second case. And this has actually been proved to be the case by oxidation with permanganate of the acids mentioned.[2]

If, on the other hand, a saturated compound becomes unsaturated, the constitution of the resulting

[1] *Lagerung der Atome im Raume*, p. 40.

[2] Kekulé and Anschütz, *Ber.* **13**, 2150; **14**, 713.

body may be foreseen in an analogous way. We may consider the isodibromosuccinic acid [1] which is prepared by the addition of bromine to maleïc acid, and hence has the formula:

$$\begin{array}{c} \text{Br} \\ \text{HCCO}_2\text{H} \\ \text{HCCO}_2\text{H} \\ \text{Br} \end{array}.$$

Let us abstract hydrobromic acid, writing the above formula in a slightly different way, in order the better to follow the result:

$$\begin{array}{l} \quad\;\; \text{H} \\ \text{CO}_2\text{HCBr} \\ \quad\;\; \text{HCCO}_2\text{H} \\ \quad\;\; \text{Br} \end{array};$$

it is then clear that we shall obtain bromofumaric acid:

$$\begin{array}{l} \text{CO}_2\text{HCBr} \\ \qquad\;\; \| \\ \qquad \text{HCCO}_2\text{H} \end{array}$$

The result is noteworthy. By addition of bromine, and subsequent splitting off of hydrobromic acid, one passes from the maleïc to the fumaric series. This transition is perfectly general and has been uniformly confirmed by observation.

Of course, as in other cases where molecular structure is to be determined, this reasoning, which is based on the stability of a molecule undergoing partial rearrangement, encounters facts apparently contradictory. Of these some, as in the above observation by Bandrowski, have been explained by

[1] *Études de dyn. chim.* p. 100.

the discovery of a secondary change. For other cases the explanation is yet to be found. We may mention, as of special interest, the conversion of the isodibromosuccinic acid, made from maleïc acid into racemic acid,[1] and that of right-handed tartaric acid into chlorofumaric acid[2] (by the action of phosphorus pentachloride[3]).

Such objections, to the number of forty-six, have been recently collected in the above-mentioned paper by Michael. But their value as a means of judging what has just been said is considerably diminished by the two following observations :

1. All the objections amount to this, that instead of the product to be expected, another results *which is more stable under the conditions of the experiment*, but in such a case a secondary transformation, masking the main result, is always possible, even in cases where this secondary action cannot be directly realised, for we have to take into account the *status nascens*.

2. All the objections refer to halogen derivatives. Now, the experiments with active compounds mentioned on p. 49, and also the reactions on p. 67, show that when *e.g.* dichlorosuccinic acid is formed from tartaric, phenylbrom- and chlor-acetic acid from malic acid, a transformation occurs. Here, too,

[1] Anschütz, *Ann. Chem. Pharm.* **226**, 191; V. Meyer, *Ber.* **21**, 264.

[2] Kauder, *J. prakt. Chem.* **31**, 33; Perkin, *J. Chem. Soc.* 1888, 645.

[3] This substance has, however, a peculiar property of reversing the position of groups in a molecule. See *ante*, p. 47.

the chlorine compounds once formed are stable, as is shown by the fact that activity is possible (p. 24); during their formation, however, transformation occurs. The practical conclusion to be drawn from Michael's work amounts to this, that in the cases investigated by him transformation easily takes place, and this is always to be expected where halogens are concerned; proof 'for' or 'against' the views above stated is therefore to be sought in cases where halogens are as far as possible excluded. Fischer in the cases mentioned (p. 82) has done this with most favourable results.

Let us now consider the mutual influence of the groups forming the molecule, so far as this can contribute to a determination of the structure of the isomers.

First, there is the question of stability. Just as our theory explained the perfectly equal stability of the two isomers of opposite activity by the absolute identity in the dimensions of the molecule, so it foresees that in general the unsaturated isomers will differ in stability, because it assumes a difference in their analogous dimensions.

Of the formulæ:

$$\begin{array}{c} R_1CR_2 \\ \| \\ R_3CR_4 \end{array} \qquad \begin{array}{c} R_1CR_2 \\ \| \\ R_4CR_3 \end{array}$$

it may, generally speaking, be maintained that, say, the second represents the more stable modification if there is reason to suppose that R_1 exerts a stronger attraction on R_4 and R_2 on R_3. Thus, for fumaric acid, stable in comparison with the isomeric maleïc acid, the formula :

$$\begin{array}{r} HCCO_2H \\ CO_2HCH \end{array}$$

seems justified.

Apart from the difficulty of comparing these attractions, we have here to take the temperature into account. Since a rise in temperature is generally opposed to the ordinary action of chemical affinities, it may happen that at a given temperature, possibly at the ordinary temperature, a transformation occurs in the sense opposed to that expected, the latter occurring only at lower temperatures. The absolute criterion of stability is, therefore, not the transformation at a given temperature, but the larger heat of formation. As is well known, on lowering the temperature the isomer with the greater heat of formation will always predominate.

Moreover there are reactions which enable us to judge as to the distance of two groups in a molecule. If in one isomer two of these groups easily undergo a simultaneous conversion, while in the other the opposite takes place, we may assume that these groups are nearer together in the first case. For example, maleïc acid readily forms an anhydride through the interaction of its two carboxyl groups,[1]

[1] Substitution of a methyl group for one or more of the hydrogens attached to carbon in maleïc acid facilitates the closing of the ring, —formation of an anhydride. The same thing is observed in the case of succinic and glutaric acids. In other cases, however, the

and is thus distinguished in a very striking way from the isomeric fumaric acid. The former, therefore, has the formula

$$\begin{array}{c} CO_2HCH \\ \parallel \\ CO_2HCH \end{array},$$

in which the two carboxyls are near together.

presence of methyl prevents the ring formation. Sometimes, indeed, in a compound containing several methyl groups, it is easier to bring about a molecular rearrangement than a simple ring formation. Thus, instead of

$$\begin{array}{c} CH_3 \\ CH_3 \cdot C \diagdown \\ | \quad O \\ CH_3 \cdot C \diagup \\ CH_3 \\ \text{I} \end{array}$$

we may obtain

$$\begin{array}{c} CH_3 \\ CH_3 \cdot C \cdot CH_3 \\ CH_3 \cdot C{:}O \\ \text{II} \end{array}$$

To account for such apparently irreconcilable observations Bischoff has applied his 'dynamic hypothesis,' according to which those configurations are the most favoured in which the components can oscillate most freely. Now, like atoms will have like paths of oscillation, and will therefore be the most prone to collide; in a favoured configuration, then, they must be far removed from one another. Hence configuration II above, in which this condition is fulfilled as regards the methyl groups, is more stable than configuration I.

Where, on the other hand, the methyl groups cause closure of the ring (*e.g.* pyrocinchonic acid, $\begin{array}{c} CH_3CCO \diagdown \\ \parallel \quad O \\ CH_3CCO \diagup \end{array}$) it is again their effort to gain room for their oscillations which causes them to crowd together the hydroxyl groups, so that expulsion of water with ring formation follows. For an account of the dynamic hypothesis, see Bischoff and Walden, *Handbuch der Stereochemie*, Frankfurt a. M. 1893–94. *Bechhold.—Tr.*

Also the formation of certain bodies may help to make clear these relationships. Thus it is plain that the closed chains occurring in benzene, cinchomeronic acid, pyromuconic acid, and pyrrol, approximate to the arrangement of the four carbon atoms in maleïc acid, and differ from the arrangement in the isomeric compound. In fact, in energetic decompositions it is maleïc acid (or its derivatives) which results in such cases.[1]

There is, finally, another, though a less direct, proof of the neighbouring position of the carboxyls in maleïc acid. This acid is the stronger: its dissociation constant is 1·17, that of its isomer only 0·093.[2] It is uniformly observed that this constant is raised by the neighbourhood of a negative group.

[1] Kekulé and Strecker, *Ann. Chem. Pharm.* **223**, 170; Hill, *Ber.* **13**, 734; Bischoff and Rach, *Ann. Chem. Pharm.* **234**, 86; Ciamician and Silber, *Ber.* **20**, 2594.

[2] Ostwald, *Zeitschr. physik. Chem.* **111**, 380.

CHAPTER VI

RING FORMATION

THE chapter devoted to ring formation in the original pamphlet was omitted in the first German edition, for at that time the isomerism of v. Baeyer's hydro- and isohydro-mellitic acids was the only case in point. Since then, however, this branch of the subject has, especially through v. Baeyer's researches, gained so much in extent and interest that an approximately systematic treatment of the whole is possible. We observe that here too the historical development has kept pace with the complexity of the problem. After the methane derivatives had been dealt with, came the ethylene and finally the polymethylene compounds.

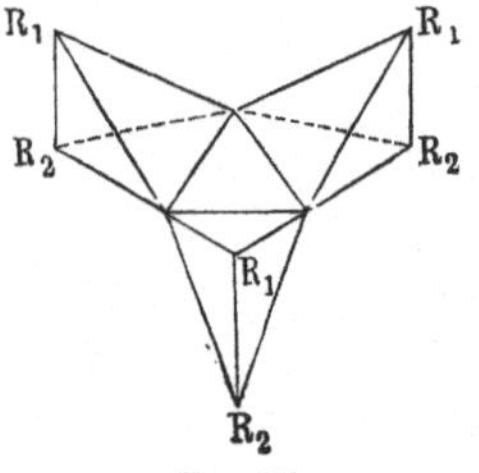

R_1 R_1 R_2 R_2 R_1 R_2

FIG. 15.

Rings of three members. Tri- and trithio-methylene.—Starting from the tetrahedral grouping, I developed, in the pamphlet referred to, the annexed configuration for the trimethylene derivatives, remarking that a transposition of the two groups R_1 and R_2, which are attached to the same carbon, would bring about an isomerism approximating to that of fumaric and maleïc acids, *i.e.* to that of dimethylene derivatives.

Since then two isomeric trimethylene dicarboxylic acids,[1] $CH_2.CHCO_2H.CHCO_2H$, and three isomeric phenyltrimethylenedicarboxylic acids,[2]

$$CHC_6H_5.CHCO_2H.CHCO_2H,$$

have in fact been discovered.

To render the discussion clearer the scheme given above may be transformed in a way readily intelligible, and the isomerism possible in the case of the trimethylenecarbonic acids may be represented in the following way:

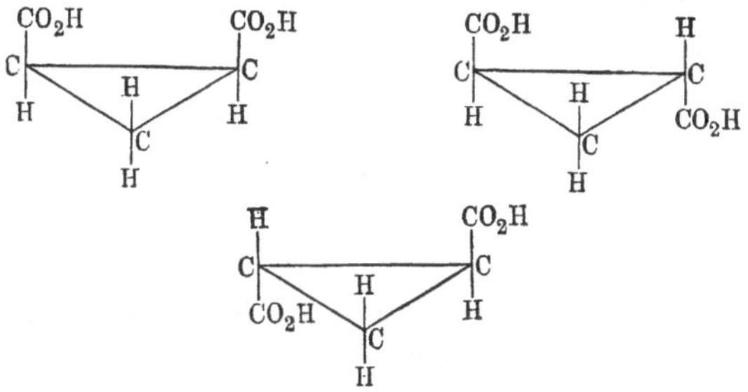

CO_2H CO_2H CO_2H H
C C C C
H H H H CO_2H
C C
H H

H CO_2H
C C
H
CO_2H H
C
H

Thus we have three possibilities, of which the second and third are non-superposable images, and must therefore possess opposite activity. Of the two known isomers, it is possible then that one may be divisible. If now a methylenehydrogen be replaced by phenyl, as in phenyltrimethylenecarboxylic acid, the first scheme leads evidently to two possibilities according as phenyl is placed above or below; the second and third schemes give, in this case, only a

[1] Buchner, *Ber.* **23**, 702.

[2] Buchner and Dessauer, *ibid.* **25**, 1148.

single isomer each, and these also are mirror images of each other. Of the three isomers found, then, one should be divisible.

Thus, although the isomerism is not analogous to that of fumaric and maleïc acid, but rather, in the first case at least, to that of inactive tartaric acid and racemic acid, it must be remembered that the first kind of isomerism is to be expected, as the figure indicates, only when all the methylene groups have undergone similar substitution. Such derivatives have not been prepared from methylene; but in the case of trithiomethylene they have been thoroughly investigated, and in accordance with the above figure they may be represented thus:

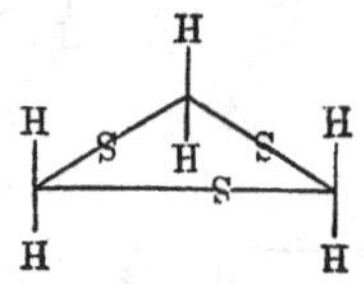

And Baumann and Fromm[1] have been led by their work on the trebly polymerised thioaldehydes and thioacetones, which probably have the constitution

$$\begin{array}{ccc} (R_1R_2)C & \text{—S—} & C(R_1R_2) \\ \diagdown & & \diagup \\ S & & S \\ & \diagdown\ \diagup & \\ & C & \\ & (R_1R_2) & \end{array}$$

to the following conclusions:

1. When the groups R_1 and R_2 are alike, as in trithiomethylene (from methylaldehyde) and trithiodimethylmethylene (from acetone), no isomerism occurs.

[1] *Ber.* **24**, 1419.

2. Two isomers occur when the groups are different, as in the aldehyde thio-derivatives of acetyl, benzoyl, anisyl, methylsalicyl, isobutylsalicyl, and cinnamyl.[1]

The observed isomers would in this case exactly correspond to the configurations given in my first pamphlet, of which one is reproduced above (fig. 15), and the other here (fig. 16). The main point to notice is that here isomerism exists without asymmetry; that is, as with fumaric and maleïc acids, no division is to be expected. The plane of symmetry lacking in the second and third trimethylenedicarboxylic acids represented above is here present.[2]

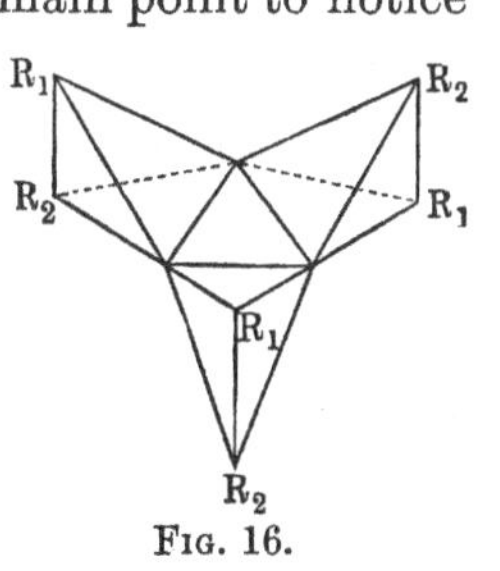

FIG. 16.

[1] But later researches show that for substituted aromatic aldehydes this holds good only when the substituting group is positive or indifferent. When it is negative no isomerism is observed. Thus there exists only one tri-thio derivative of the following: methoxybenzaldehyde, benzoylmethoxybenzaldehyde, methylmetoxybenzaldehyde, paroxybenzaldehyde, benzoylparoxybenzaldehyde, vanillin (but methylvanillin yields the isomers), benzoylvanillin, gentisinaldehyde, metanitro-, anis-, and cumin-aldehyde, dinitroanisaldehyde (Wörner, *Ber.* 29, 139).

[2] It is evident that if the radicals R_1R_2 are in the plane of the trimethylene ring there must be three inactive trimethylenedicarboxylic acids, of the formula $CH_2.CHCOOH.CHCOOH$.

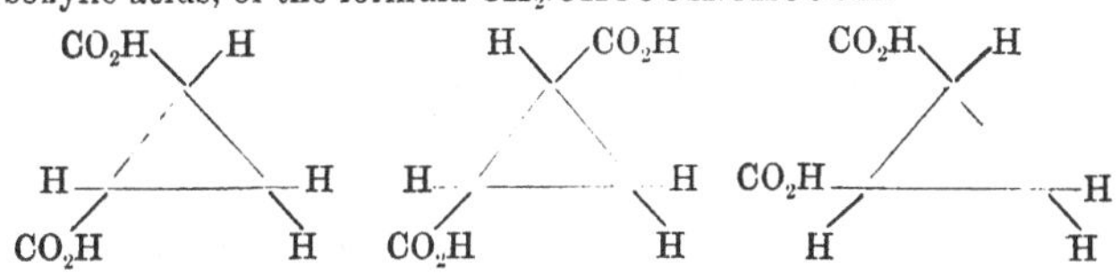

Whereas we have seen that according to the tetrahedron hypothesis there can be only two inactive isomers, of which one should be

Rings of four members. Tetramethylene derivatives.—Here too, especially through Liebermann's[1] investigation of the truxillic acids, cases of isomerism have been discovered which may well be classed with those above mentioned. The acids named, which from their transformation must be considered as dicinnamic acids, and from their saturated character as tetramethylene derivatives, correspond to the two formulæ:

$$\begin{array}{ccc} C_6H_5CH—CHCO_2H & \qquad & C_6H_5CH—CHCO_2H \\ \quad | \qquad\quad | & & \qquad | \qquad\quad | \\ C_6H_5CH—CHCO_2H & & CO_2HCH—CHC_6H_5 \end{array}$$

and have been obtained in four, if not in five, isomeric forms.

The first formula alone presents the following possibilities:

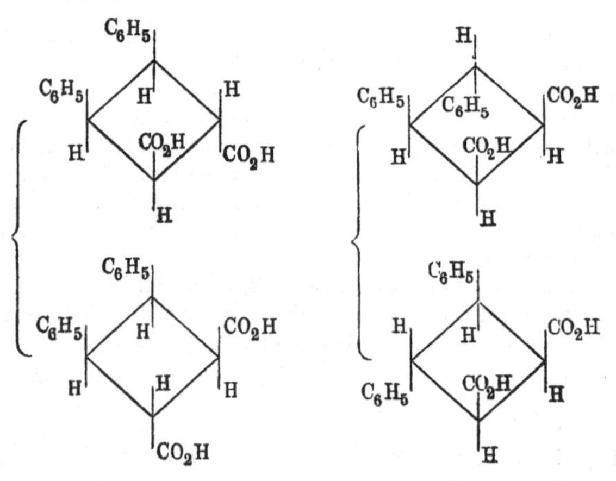

divisible. Buchner's later results are all in favour of this hypothesis. He has also prepared the stereomers of the tri- and tetra-carboxylic acids (*Ann.* **286**, 197).

[1] *Ber.* **23**, 2516.

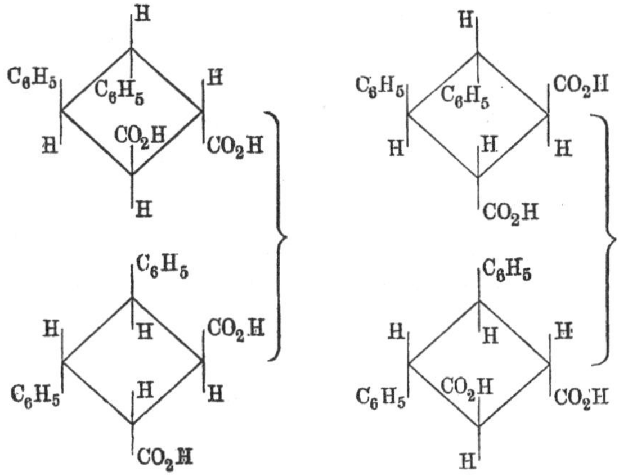

of which the figures bracketed together are mirror images without symmetry. We have to expect, then, six isomers, of which four are divisible—that is, altogether, ten different isomers.

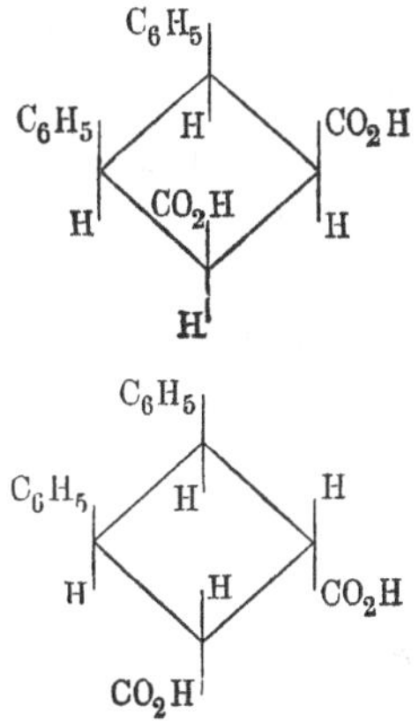

Next we have to mention the so-called α-γ-diacipiperazines,[1] substitution products and homologues of the compound

$$C_6H_5N\begin{matrix}COCH_2\\ \\CH_2CO\end{matrix}NC_6H_5.$$

Disregarding for the present the possibility, which will be discussed later, that the nitrogen may cause

[1] Bischoff, *Ber.* **25**, 2950.

isomerism, there is none to be expected in the derivatives investigated, which belong to the type

$$\begin{array}{ccc} & COCH_2 & \\ XN & & NX \\ & CH_2CO & \end{array}$$

and, in fact, for

$X = C_6H_5$, $C_6H_4CH_3$ (p and o), $C_{10}H_7$ (α and β), none was found; nor even when the X groups were different (C_6H_5 and C_7H_7).

But if in the two methylene groups a hydrogen is replaced by CH_3 or C_2H_5, giving the type

$$\begin{array}{ccc} & CO\mathit{C}HR & \\ XN & & NX, \\ & \mathit{C}HRCO & \end{array}$$

isomerism results, and may be considered as due to the two asymmetric carbon atoms.

The isomers found for

$X = C_6H_5$, $C_6H_4CH_3$ (p and o), $C_{10}H_7$ (α and β), which have been prepared in ten cases, would correspond to the two possible racemic mixtures:

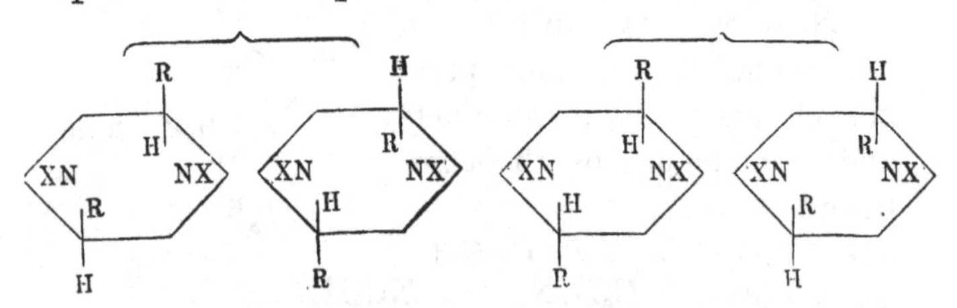

Rings of six members. Hexamethylene derivatives. In the case of the derivatives of hexamethylene the observations are no less convincing. While among the monosubstituted hexamethylene derivatives, such as hexahydrobenzoic acid, no case of isomerism is as yet known, it is otherwise with the products which have undergone several substitutions; and we owe to Baeyer's[1] researches a knowledge approximately complete of all the details in at least one case. This is the hydroterephthalic acid, especially the hexahydro derivative, $C_6H_{10}(CO_2H)_2$ 1, 4. Of this two modifications have been discovered which correspond to the cases foreseen:

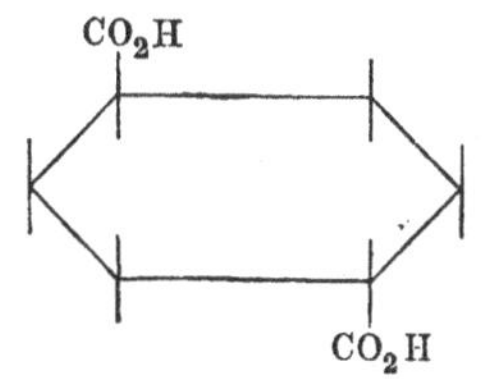

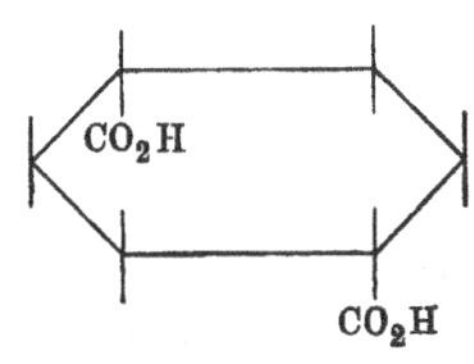

In these figures the twelve groups linked in pairs with carbon or the atoms of the ring form the corners of a hexagonal prism the edges of which are indicated by the vertical lines. For simplicity the carbon atoms of the ring, as well as the hydrogen atoms remaining unsubstituted, are again omitted in the figure.

Here, too, the hexahydrophthalic acids,

$$C_6H_{10}(CO_2H)_2 \ 1, 2,$$

which likewise occur in two modifications, may be mentioned:

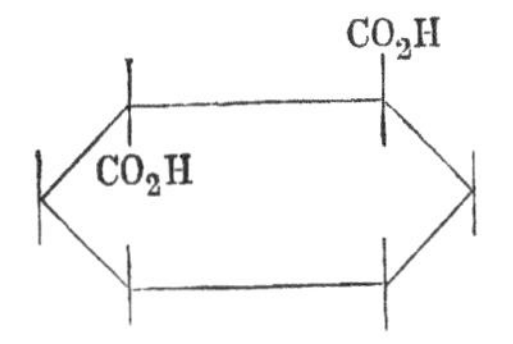

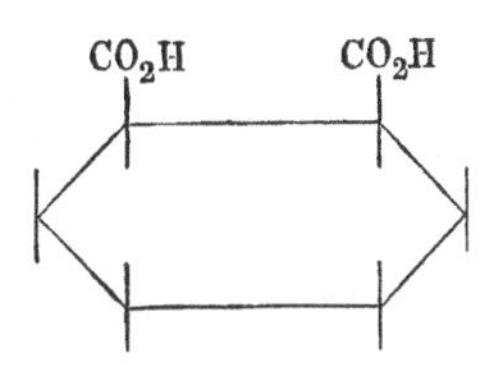

[1] *Ann. Chem. Pharm.* **245**, 103; **251**, 258; **258**, 1, 145.

In this case, however, it must be added that the first arrangement corresponds to an enantiomorphous form, and that accordingly a division into two active isomers must be possible.[1] The two compounds obtained would be comparable to racemic acid and inactive tartaric acid, for they contain two asymmetric carbon atoms in a symmetrical formula. The recent discovery of the two isomeric hexaisophthalic acids by Perkin[2] has supplemented the above.

Finally, we may mention the *α*- and *β*-tetrahydroterpenes, the terpines, and the pinenedihydrochlorides and bromides,[3] which perhaps correspond in structure to the hexahydroterephthalic acids.

If more than two hydrogen atoms are substituted in hexamethylene, the number of possible isomers will be increased. On this point, however, the number of researches is limited: as cases of treble substitution, only the bromides of tetrahydrobenzoic acid[4] can be mentioned.

It is only when we come to the sixfold substitution products that the number of compounds investigated increases; hydro- and isohydro-mellitic acids, $C_6H_6(CO_2H)_6$, having been the first cases

[1] Such a division has since been effected in the case of the analogous compound cis-trans-hexahydroquinolic acid,

$$\begin{array}{l} \qquad\;\; H_2\;\; H \\ H_2C-C-\mathit{C}COOH \\ \qquad\; | \qquad\;\; | \\ H_2C-N-\mathit{C}H \\ \qquad\;\; H\;\;\; COOH \end{array}$$

(Besthorn, *Ber.* **28**, 3153).

[2] *J. Chem. Soc.* 1891, 814.

[3] Beilstein, 2nd ed. 1, 182; Baeyer, *Ber.* **26**, 2861.

[4] *Ber.* **24**, 1867.

investigated.[1] With regard to the isomeric hexachlorbenzenes, $C_6H_6Cl_6$, on which so much work has been done recently, since the identity of their molecular weights has been proved [2] the assumption of a structural difference is scarcely tenable, and accordingly Friedel [3] and Matthews [4] resorted to the stereochemical explanation. This is included in what has been said above.

Activity among hexamethylene derivatives. Inosite. It has already been repeatedly stated that among polymethylene derivatives optical activity is to be expected. And it has been proved to exist in several cases, viz.:

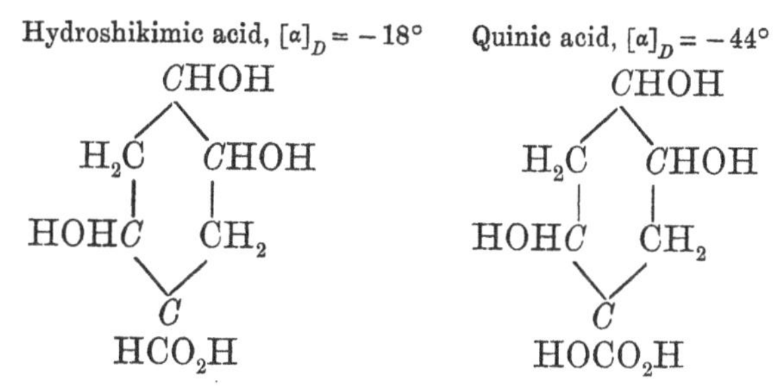

This might have been at once foreseen as a consequence of the asymmetric carbon atoms which are evidently present. The case of inosite, however, demands special attention:

[1] Baeyer, *Ann. Chem. Pharm.* Suppl. **7**, 43.

[2] Paterno, *Gazz. Chim.* **19**, 195.

[3] *Bull. Soc. Chim.* [3], **5**, 130.

[4] *Chem. Soc. J.* 1891, 165; the same author has recently prepared two isomeric chlorbenzenehexachlorides. Recently, too, Orndorff and Howells (*Am. Chem. J.* **18**, 312) announce the discovery of stereomerism in the case of hexabromobenzene.

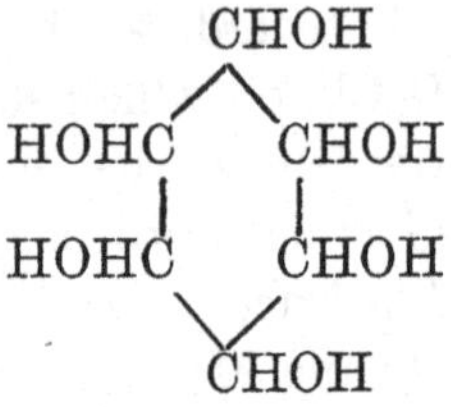

CHOH

HOHC CHOH

HOHC CHOH

CHOH

Here the absence of symmetry due to the presence of the asymmetric carbon is not evident, or at least not sufficiently so. It shows itself in these cases only on consideration of the scheme developed as above, which is therefore applied here. The ordinary long-known inosite is inactive and indivisible (p. 46), and therefore possesses a symmetrical constitution :

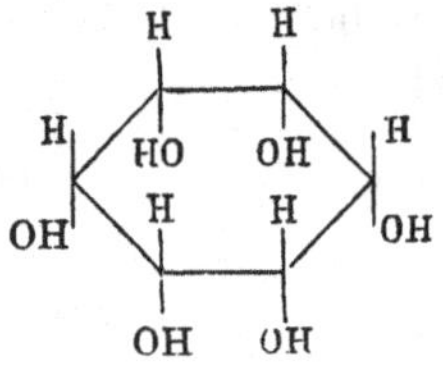

H H

H HO OH H

OH H H OH

OH OH

The right and left inosites, $[\alpha]_D = 65°$, prepared by Maquenne,[1] the first from pinite and β-pinite, the second from quebrachite—that is to say, both from isomeric methylinosites, $C_6H_6(OH)_5OCH_3$—will then correspond to the asymmetric mirror images, such as, *e.g.*:

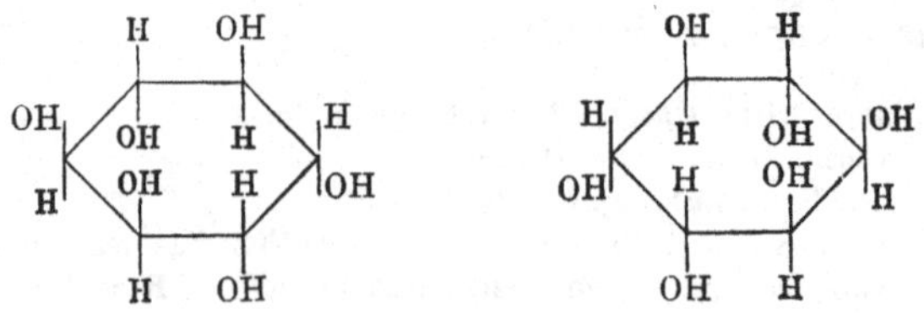

H OH OH H

OH OH H H H H OH OH

H OH H OH OH H OH H

H OH OH H

[1] *Ann. Chim. Phys.* [6], **22**, 264; *Compt. Rend.* **109**, 812.

These diagrams explain also how it is that the same inosite can result from isomeric methyl derivatives, pinite and β-pinite. It may be observed parenthetically that isomers in considerable variety are to be foreseen here, and are perhaps to be found in scyllite, in phenose, &c.

Finally, hexahydro-o-toluylic acid,

$$\begin{matrix} & H \\ < & CH_3 \\ & H \\ < & CO_2H \end{matrix}$$

has been obtained by Goodwin and Perkin jun.[1] in two modifications representing the *cis*[2] and *trans*[2] configurations.

Tetrahydrobenzene derivatives.—Starting from the derivatives of hexamethylene, to which the application of stereochemical conceptions is simple, we gradually arrive—using Baeyer's investigations on the tetra- and di-hydrides of terephthalic acid—at the complicated state of things presented by the benzene nucleus.

Thus, if in the isomeric tetrahydro derivatives we assume a double bond, it is easy to see that the following two forms must exist:

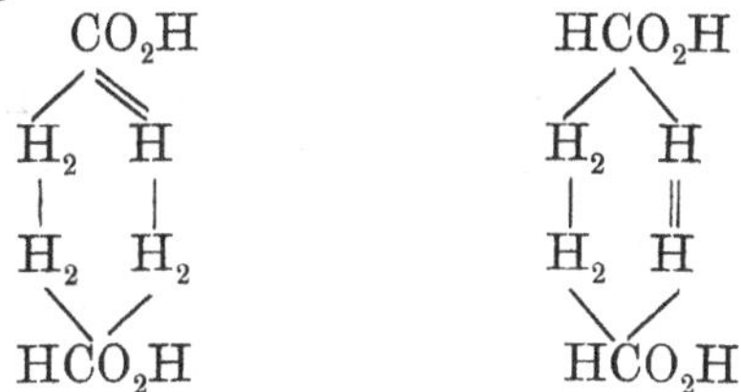

[1] *J. Chem. Soc.* 1895, i. 119.

[2] These terms were introduced by Baeyer to distinguish the isomer in which the substituents are on the same side, from that in which they are on opposite sides of the ring.

Of the latter formula, the two following isomers are to be expected :

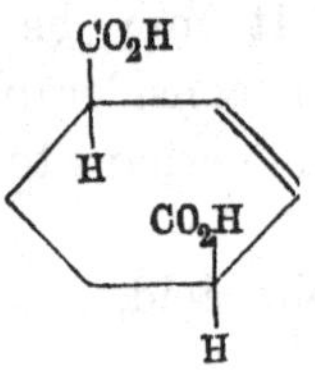

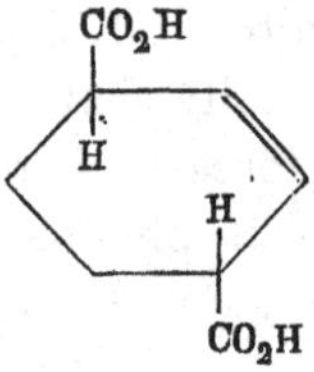

CO_2H H CO_2H H CO_2H H   H CO_2H

And, in fact, three tetrahydroterephthalic acids are known. Again, we observe that the last diagram is enantiomorphous, and therefore it is to be expected that two of the acids should stand to one another in the relation of inactive tartaric acid and racemic acid ; for the formula to which they correspond contains two asymmetric carbon atoms, the constitution of the whole being symmetrical.

Dihydrobenzene derivatives.—Here, too, we find theory and observation in accord. Assuming two double bonds, the following four structural formulæ are to be expected :

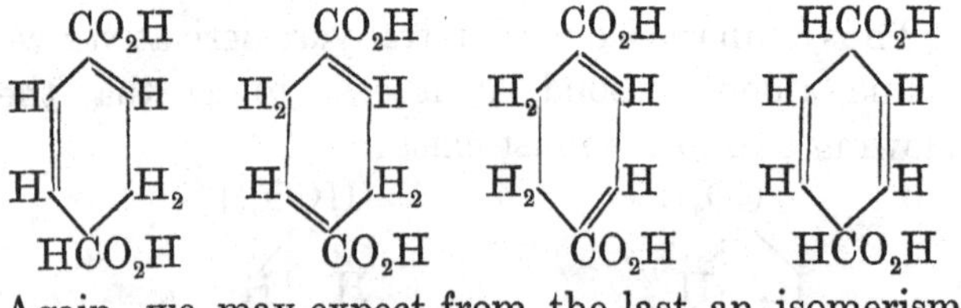

CO_2H	CO_2H	CO_2H	HCO_2H
H H	H_2 H	H_2 H	H H
H H_2	H H_2	H_2 H	H H
HCO_2H	CO_2H	CO_2H	HCO_2H

Again, we may expect from the last an isomerism,

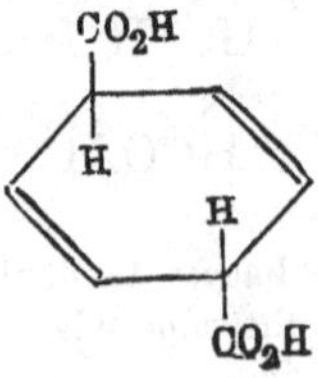

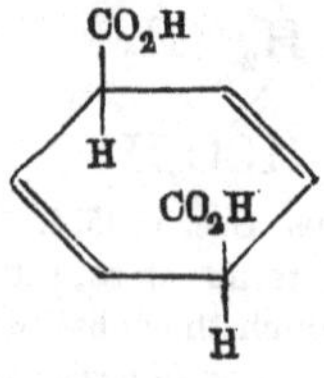

CO_2H H   H CO_2H CO_2H H CO_2H H

which probably corresponds to the fumaric-maleïc isomerism. In fact, Baeyer has described five dihydroterephthalic acids.

Benzene derivatives.—When finally we come to the derivatives of non-hydrogenated benzene, the ethylene character is lacking and with it the condition determining isomerism; the acetylene character is now assumed, and thus all ground for stereochemical speculations vanishes. This point must be specially emphasised. Everything forces on us the conclusion that in these non-hydrogenated derivatives of benzene rotatory power is lacking, unless the side chain contains an asymmetric atom. This assumption is based in the first place on the fact that the numerous benzene derivatives occurring in nature, such as salicylic aldehyde, vanillin, cumarin, are without exception inactive; and in the next place all attempts at 'doubling' have been in vain. Le Bel[1] has made such attempts with orthotoluidine; Lewkowitsch[2] with β-meta-homosalicylic acid,

$$C_6H_3(CH_3)(CO_2H)(OH)(1, 2, 3,),$$

with β-ortho-homomethoxybenzoic acid,

$$C_6H_3(OH)(CH_3)(CO_2H)(1, 2, 3,),$$

and with methoxytoluylic acid,

$$C_6H_3(OCH_3)(CH_3)(CO_2H)(1, 2, 3);$$

and V. Meyer and F. Lühn[3] with nitro- and formyl-thymotic acids,

$$C_6H.OH.COOH.CH_3.C_3H_7.NO_2$$

[1] *Bull. Soc. Chim.* **38**, 98.

[2] *Chem. Soc. J.* 1888, p. 791; see also *Ber.* **16**, 1576.

[3] *Ber.* **28**, 2795

and $C_6H.OH.COOH.CH_3C_3H_7.CHO$, without effecting a division.[1] Now, only those benzene formulæ which contain the carbon and hydrogen atoms in one plane can be free from enantiomorphism, which in the prism formula, indeed, would manifest itself even in the bisubstitution products.

Objection.—The ring-linkage has here been treated in such a way that, starting from methylene derivatives, we advanced gradually to benzene. In this way, however, a difficulty is concealed which must now be mentioned. The construction of the configurations by means of models does very well in the case of methylene derivatives, as is shown by the figures sketched. In constructing benzene, however, we encounter, if we take Kekulé's doctrine as a basis, the well-known difference[2] between 1, 2, and 1, 6; while, with Ladenburg's prism, activity is to be expected even in bisubstituted products. This objection is met, however, if we consider the tetrahedral grouping as only the cause of the final arrangement of the atoms, which, in benzene, adopting a plane arrangement, would be

```
      H     H
        C C
   H  C     C  H
        C C
      H     H
```

The tetrahedra, then, are to be considered as the cause of the grouping, not as anything really present.

[1] According to Rügheimer, however, active m-methylic-p-oxybenzoic acid possibly exists (*Ber.* **29**, 1967).

[2] See Graebe, *Ber.* **29**, 2802.

Stability of the ring formation.—Let us apply, finally, the tetrahedron theory from a point of view somewhat different. It is now some years since Victor Meyer,[1] on the occasion of a general review of the essential properties of carbon compounds, referred to the peculiar readiness of this element to form closed chains consisting of six atoms, and to the extraordinary stability of these compounds. In view of the difficulty of obtaining closed chains with, for example, three atoms—such bodies were at that time not even known—this was all the more remarkable. And Victor Meyer urged with justice that such an essential property should be deducible as an immediate consequence from clearer views as to the structure of organic compounds.

I took this occasion to point out [2] that the new tetrahedron theory was capable of explaining this peculiarity up to a certain point. However, these considerations attracted no further notice ; and there would be no reason to produce here those rather tentative remarks, but that recently Baeyer,[3] Wunderlich,[4] and Wislicenus,[5] each from his own standpoint, have developed ideas of a perfectly analogous nature. It seems therefore desirable to repeat here the observations referred to.

For the various observations which have led to analogous conceptions on the part of the chemists

[1] *Ann. Chem. Pharm.* **180**, 192.
[2] *Maandblad voor Natuurwetenschappen*, **6**, 150.
[3] *Ber.* **18**, 2278.
[4] *Konfiguration organischer Moleküle*, Würzburg, 1886.
[5] *Abh. der Königl. Sächs. Akad.* 1887, 57.

named, possess one characteristic in common. Bodies containing a chain of several connected carbon atoms are sometimes capable of remarkable transformations, arising from a preference for interaction between distant groups. To give an example. Among the oxybutyric acids it is precisely that one which most readily forms a lactone which has the carboxyl and hydroxyl groups apparently furthest removed from one another, namely, γ-oxybutyric acid, $CO_2HCH_2CH_2OH$. The anhydride results, then, from the interaction of the two outermost groups with loss of water. This phenomenon is general; the γ-oxy-acids, which have three carbon atoms between hydroxyl and carboxyl, are always those which display a special tendency to lactone formation.

Now, our theory, so far from seeing any difficulty in the interaction of groups attached to the carbon atoms of a long chain, finds here, if not a direct confirmation, at least the indication of such. Let us represent the grouping of several carbon atoms according to our views. The first carbon atom, C_1, with the two groups it connects, will be at the corners of an isosceles triangle, the angle at A being, according to the dimensions of the tetrahedron, 35°. The second carbon atom, C_2, with the connected C_1 and C_2, will be arranged in an absolutely identical fashion. The same holds for a third

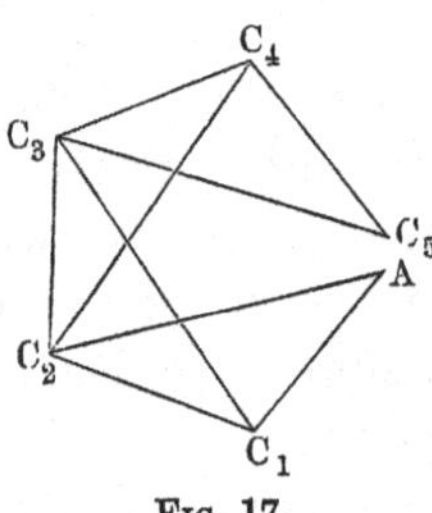

Fig. 17.

atom, C_3, for a fourth, C_4, and so on. Now, it is plain that the distances A C_2, A C_3, A C_4, which represent the distances of the groups connected with the first, with the first and second, and with the first and third atoms, do not continually increase.

On the contrary, since the ratio of these distances is expressed by

$$\sin 2A : \sin 3A : \sin 4A : \sin 5A = 1 : 1\cdot 02 : 0\cdot 67 : 0\cdot 07,$$

there must ensue, as the figure also shows, a considerable decrease in the distances in question.

After these general considerations, let us pass on to the discussion of details.

With regard to Baeyer's[1] views, we note first that this author assumes in the closed-chain polymethylenes a symmetrical arrangement of the carbon atoms, and compares the angle which two carbon atoms make with a third connected with them, with the angle $C_2\,C_1\,A$ of fig. 17. Now, according as we have to do with hexa-, penta-, tetra-, tri-, or dimethylene, this angle is 120°, 106°, 90°, 60°, or 0°, while the angle $C_2\,C_1\,A$ of fig. 17 is about 109°. The difference is, then, 11°, 3°, 19°, 49°, and 109° respectively, and in this difference the author sees an approximate expression of the tendency to saturation. In support of this view may be mentioned the extraordinary difficulty of saturating hexa- and tetra-methylene; whereas trimethylene unites with bromine, though not with hydrobromic acid. In the case of dimethylene even the action of iodine suffices to bring about saturation.

[1] *Ber.* **18**, 2278.

The benzene derivatives admit of similar treatment,[1] which, however, is influenced by the fact that the relative position of the six carbon atoms is here not quite settled. We assume Kekulé's hypothesis, according to which double and single bonds alternate, and compare with benzene those analogous closed chains of five to eight carbon atoms, which may be assumed from the valence of carbon, namely,

$(CH)_4$, $(CH)_4CH_2$, $(CH)_6$, $(CH)_6CH_2$, and $(CH)_8$.

To this end let us place side by side the sums of the angles which our theory requires when two carbon atoms are joined to a third (about 109° in the case of a single, and 125° in the case of a double bond), with the sums of the angles of a closed polyhedron :

Formula	Sum of the angles	Polyhedron angles	Difference
$(CH)_4$ . .	$4 \times 125 = 500$	300	140
$(CH)_4CH_2$.	$4 \times 125 + 109 = 609$	540	69
$(CH)_6$ . .	$6 \times 125 = 750$	720	30
$(CH)_6CH_2$.	$6 \times 125 + 109 = 859$	900	−41
$(CH)_8$ . .	$8 \times 125 = 1000$	1080	−80

We see that, in fact, the greatest approximation occurs in the case of benzene, which accounts for the stability of this substance as well as for the fact that up to the present the others have not been prepared.

[1] Wunderlich, *Konfiguration organischer Moleküle*; van 't Hoff *Maandblad voor Natuurwetenschappen*, 7, 150.

CHAPTER VII

NUMERICAL VALUE OF THE ROTATORY POWER

WHEREAS thus far we have spoken only of the absence or presence of rotatory power, we have now to do with the magnitude of the rotation. It is already a considerable time since such determinations began to be made, and (as the expression of the quantitative relation) the so-called molecular rotation was chosen—that is, the specific rotation, a,[1] multiplied by the molecular weight (and for shortness divided by 100). The chief results so obtained are, firstly, the statement of Mulder, Krecke, and Thomson[2] that the molecular rotations within certain groups of substances bear a simple ratio to one another; and, secondly, the observation of Oudemans and Landolt that different salts of the same active base or acid in dilute aqueous solution possess the same molecular rotation. Such considerations have gained a new interest for stereochemistry since Guye[3] and Crum Brown[4] attempted

[1] Rotation caused by 1 decim., the substance being supposed present in this column with the density one.

[2] *Zeitschr. f. Chemie*, 1868, **58**; *Zeitschr. f. prakt. Chem.* 1872, **5** 6; *Ber.* 1880, 1881.

[3] *Thèses*, 1891; *Ann. Chim. et Phys.* [6], **25**, 145.

[4] *Proc. Roy. Soc. Edinb.* **17**, 181.

to connect the magnitude of the rotation with the nature of the groups attached to the asymmetric carbon atom; accordingly the facts bearing on the question are here given in detail.

I. Comparison of the Numerical Results. Necessity of an Examination in Dilute Solution and of taking into account the Molecular Weight

It was *à priori* certain that the relation between the groups attached to the asymmetric carbon and the rotation must be such that when two groups become identical the rotation vanishes; but in attempting to go beyond this we are at once met by the difficulty that the magnitude of the rotation depends on the wave-length of the light, on the solvent, and on the temperature. The first thing is, then, to determine the conditions in which comparable numbers may be obtained.

And here it seems most essential to avail ourselves of the light thrown on the subject by the new conception of the nature of solutions.

It is certainly inadmissible to use simply the figures obtained by an examination of the substance without special precautions, because the size of the molecule is then uncertain, and the magnitude of the rotation seems to be specially influenced by every change of constitution. In this connection it is important to remember the fact recently discovered by Ramsay,[1] that, of fifty-seven liquids examined, no

[1] *Chem. Soc. J.* 1893, 1098.

less than twenty-one possessed double molecules, among them the alcohols, acids, nitro-ethane, aceto-nitrile, and acetone. Another objection is that the rotation is generally influenced by the solvent, and, indeed, by every solvent differently, perhaps in consequence of the four groups attached to carbon being differently attracted. If the substance be used alone, without solvent, its own molecules may be supposed to exert a similar influence, an influence displayed most prominently in the formation of crystals, and which, in the case of strychnine sulphate, *e.g.*, leads to the almost complete annihilation of the rotation.

The objections mentioned disappear completely only when the substance is examined in the state of rarefied gas. As this is impracticable we are driven to adopt some other means, and thus arrive naturally at the state of dilute solution. It is also indispensable, of course, to take into account the molecular weight, which can then easily be determined; while the comparability of the results will evidently be by far the greatest when the same solvent is chosen for the different cases.

The influence of wave-length and of temperature seems not to be important if the circumstances of each case are duly taken into account. The anomalous rotation-dispersion in the case of, say, tartaric acid in aqueous solution—which is such that the rotation changes its direction with the colour—is evidently connected with phenomena of equilibrium which affect the tartaric acid in the solution; it was also found by Biot in a mixture of right- and left-handed substance. The same holds for the great alteration in the rotation of tartaric acid when the temperature, the concentration, or the solvent is changed. All these phenomena are connected together and only make necessary a careful use of the figures obtained, but are no argument against the existence of relations between rotation and constitution in general.

II. Rotatory Power of Electrolytes. Law of Oudemans-Landolt

Active bases.—In perfect harmony with the new views of the nature of aqueous solutions—according to which electrolytes undergo, at a sufficient degree of dilution, a division into ions until, as Arrhenius pointed out, a limit is reached—stands Oudemans' observation concerning salts of active bases and acids. At a sufficient degree of dilution the molecular rotation of quinine, *e.g.*, is independent of the salt observed. The following table (p. 137) gives the results obtained by Oudemans[1] and also by Tykociner;[2] it gives the specific rotation $[\alpha]_D^{20}$, observed at 16° C., and calculated for the base.

It may be remarked here that the equality of rotation which Wyrouboff[3] recently showed to exist in solutions of isomorphous sulphates and selenates

[1] *Rec. des Trav. Chim. des Pays-Bas*, **1**, 18, 184.

[2] *l.c.* **1**, 144. For nicotine, Schwebel, *Ber.* **15**, 2850; Carrara, *Gazz. Chim.* **23**, [2], 593.

[3] *Compt. Rend.* **115**, 832.

TABLE I.

	ClH	NO_3H	ClO_3H	$C_2H_4O_2$	CH_2O_2	SO_4H_2	$C_2H_2O_4$	PO_4H_3	BrH	ClO_4H	$C_6H_8O_7$
Quinamine . .	+ 110	118	117	118	118	117	118	117	—	—	—
Conquinamine . .	+ 228	—	—	229	228	229	228	229	229	—	—
Quinine . . .	- 279	284	286	279	281	279	272	280	—	288	—
Quinidine . . .	+ 326	329	329	318	326	322	316	325	—	334	—
Cinchonine . .	− 259	258	263	251	259	259	254	259	256	263	—
Cinchonidine . .	− 176	178	183	174	178	180	178	180	—	183	—
Apocinchonine . . .	+ 212	213	216	206	216	213	208	214	213	218	206
Hydrochlorapocin-chonine . .	+ 227	226	231	227	229	227	225	235	225	229	223
Brucine . . .	− 35·9	—	—	35·8	36·5	34	34·1	35·5	35·5	AsO_4H_3	35·9
Strychnine . .	− 34·7	34·4	—	34	34	35	34·4	34·4	—	34	34·4
Morphine . . .	− 128	128	—	129	129	128	128	128	—	128	128
Codeine . . .	− 134	134	—	135	135	134	134	$C_3H_6O_2$	—	134	134
Nicotine . . .	+ 14·4	12·6	—	13·8	—	14·5	—	12·2	12·2	—	—

of strychnine and cinchonine, is by no means to be considered, as he says, as a consequence of a connection between rotation and isomorphism; it is simply a confirmation of Oudemans' law.

Active acids.—The same holds for the salts of active acids as Landolt found for tartaric acid, and as is proved by Table II., where the specific rotation, calculated for the acid, is given (see p. 139).

The salts of shikimic acid, with alkalies and alkaline earths, also exhibit equal rotation, according to Eykman;[1] and the same holds, according to Colson, for acetylmalic acid.[2]

Finally, it must be mentioned, with regard to the remarkably low figure obtained for the barium and calcium salts of methoxy- and ethoxy-succinic acid, that the very great influence of concentration is here to be taken into consideration. The specific rotation of the methoxybarium salt is, *e.g.*, for the percentage given:

26·1 per cent.	12·4 per cent.	5·7 per cent.	1·15 per cent.
− 14·3	− 7·4	− 2·2	+ 3·2

Evidently the limit of dissociation is not yet reached, and this is probably true also for the glycerates of the polyvalent metals. With the monovalent metals the maximum seems to be reached sooner. The glycerates were investigated in ten per cent. solution.

In these investigations the theory of electrolytic dissociation is a valuable guide; it enables the

[1] *Ber.* **26**, 1281. [2] *Compt. Rend.* **116**, 818.

TABLE II.

	Li	Na	K	NH_4	Ca	Sr	Ba	Mg	Zn	Cd
Podocarpinic acid [1] .	—	+133	134	133	—	—	—	—	—	—
Quinic acid [1] . .	—	− 48·9	48·8	47·9	48·7	48·7	46·6	47·8	51	—
Cholalic acid [2] . .	—	+ 28·6	31	—	—	—	—	—	—	—
Camphoric acid [3] .	+19·5	20·2	19·4	19·7	19·6	—	19·9	19·8	—	—
Tartaric acid [4] . .	+38·6	39·9	43	42	—	—	—	41·2	—	—
,, ,, , acid salts	+28·5	27·5	28·3	28·5	—	—	—	—	—	—
Malic acid [5] . .	−13·9	13·1	11·5	11·2	—	—	> 5	—	—	—
,, ,, , acid salts	− 8·8	8·2	8	7·7	—	—	—	—	—	—
Glyceric acid [6] . .	−22	19	22	21	16	17	16	22	31	23
Methoxysuccinic acid [7]	—	—	+ 14	15	> 6	—	> 6	—	—	—
Methoxysuccinic acid, acid salts . .	—	—	+ 29	29	—	—	—	—	—	—
Ethoxysuccinic acid .	—	—	—	+ 23	>14	—	>12	—	—	—
Ethoxysuccinic acid, acid salts . .	—	—	+ 36	+ 37	—	—	—	—	—	—
Acetylmalic acid [8] .	—	—	− 2	—	—	—	− 2	—	—	Ag
β-Oxybutyric acid [9] .	—	− 18	—	—	—	—	—	—	—	−20

[1] Oudemans, *Rec. des Trav. Chim. des Pays-Bas*, **4**, 166. [2] Hoppe-Seyler, *J. f. pr. Chem* **89**, 264, 272. [3] Hartmann, *Ber.* **21**, 221.
[4] Landolt, *ibid.* **6**, 1073. [5] Schneider, *Ann.* **207**, 284. [6] Frankland and Appleyard, *Chem. Soc. J. Trans.* 1893, 311.
[7] Purdie and Walker, *ibid.* 1893, 248. [8] Colson, *Compt. Rend.* **116**, 816. [9] Minkowsky, *Ber.* **17**, Ref. 535.

Oudemans-Landolt law to be predicted, and sees in the equality of rotation of the different salts the consequence of the existence of the same ions. Table II., then, may be condensed thus:

$[\alpha]_D$ for the ion	$CO\bar{O}(\mathit{C}HOH)_2CO\bar{O}$	43°	Diff. 14
„ „ „ „	$CO\bar{O}(\mathit{C}HOH)_2CO_2H$	29°	
„ „ „ „	$CO\bar{O}\mathit{C}HOHCH_2CO\bar{O}$	14°	
„ „ „ „	$CO\bar{O}\mathit{C}HOHCH_2CO_2H$	9°	
„ „ „ „	$CH_2OH\mathit{C}HOHCO\bar{O}$	22°	
„ „ „ „	$CO\bar{O}\mathit{C}HOCH_3CH_2CO\bar{O}$	15°	Diff. 14
„ „ „ „	$CO\bar{O}\mathit{C}HOCH_3CH_2CO_2H$	29°	
„ „ „ „	$CO\bar{O}\mathit{C}HOC_2H_5CH_2CO\bar{O}$	23°	Diff. 14
„ „ „ „	$COO\mathit{C}HOC_2H_5CH_2CO_2H$	37°	

From this we see at once that, when the rotation alters on dilution, only the values at the limit are to be taken, and doubtful cases may be decided by a determination of the conductivity—*i.e.* of the molecular weight—accompanying the observation of the polarisation. Then the objection recently brought by Frankland against Oudemans' law, based on the abnormally large rotation of tartar emetic, at once breaks down; for this salt, according to determinations of the molecular weight and to the chemical reactions, is present in solution in a form quite different from the other tartrates.[1]

Alcoholic solutions of electrolytes.—Of alcoholic solutions, at least some have been investigated. It is probable that here division into ions is not of such frequent occurrence. Also, the results vary more (for quinates,[2] *e.g.*, they lie between −9° and −40°, while in water the extremes are −43° and −49°); how-

[1] Hädrich, *Zeitschr. f. physik. Chem.* **12**, 476.

[2] Cerkez, *Compt. Rend.* **117**, 173.

ever, the hydriodide, perchlorate, and nitrate of quinamine are equal. All that is to be inferred from this is that here, as in other instances, bodies capable of undergoing division into ions often, without being actually divided, show in their physical properties an approximation to the products of division.

The influence which may be exerted by electrolytic dissociation is evident from the following conspectus of the results in alcohol and in water, which contains the limiting values obtained for various salts :

	Alcohol	Diff.	Water	Diff.
Quinamine salts . .	130 to 135	5	117 to 118	1
Conquinamine . .	200 „ 234	34	228 „ 229	1
Quinidine . . .	233 „ 255	22	322 „ 329	7
Cinchonine . . .	206 „ 240	34	258 „ 289	1
Cinchonidine . .	−114 „ −161	47	176 „ 180	4
Quinic acid salts . .	−9 „ −40	31	−43 „ −49	6
Quinine sulphate .	−212	—	−279	—
Nicotine acetate . .	− 65	—	+13·8	—

The change of sign in the case of nicotine salts (with the sulphate [1] also) is of especial interest.

III. Rotation of Imperfect Electrolytes. Organic Acids

These substances demand separate treatment because, representing as they do the transition stage between electrolytes and non-electrolytes, they exhibit—in aqueous solution at least—complicated phenomena, which, however, have already been partially accounted for. In view of the alteration of

[1] Nasini, *Gazz. Chim.* 1893, 43.

the molecular conductivity and of the lowering of the freezing-point with the concentration of their solutions, it is evident that water effects a fundamental change in their molecular structure—dissociation, in fact. The salts, especially those of strong acids and bases, show this at degrees of dilution which admit of an optical examination, and then Oudemans' law holds. With the acids this is not the case.

While, *e.g.*, the non-electrolyte sugar,[1] at a strength of from 70 to 0·2 per cent., shows a scarcely noticeable alteration of $[a]_D = 64{\cdot}5$ to $65{\cdot}2$, and for disodium tartrate[2] the rotation for concentrations (c) between 5 and 15 per cent. is expressed by

$$[a]_D^{20} = 27{\cdot}85 - 0{\cdot}17\,c \; (25{\cdot}3 \text{ to } 27),$$

for tartaric acid[3] we have

$$[a]_D^{25} = 14{\cdot}98 - 0{\cdot}1303\,c \; (8{\cdot}5 \text{ to } 14{\cdot}3)$$

between $c = 50$ and 5; while the rotation between 4·7 per cent. and 0·35 per cent. rose from 14·2 to 16·3 (at 20°). Malic acid even changes from left to right according as dilute or concentrated solutions are used.[4]

The laws which govern these complex phenomena are the following:

1. The alteration in rotation effected by change of concentration is parallel with that effected by change of temperature, dilution and rise of tempera-

[1] Schmitz, Tollens, *Ber.* 10, 1414, 1403; Přibram, *Sitz.-Ber. preuss. Akad.* 1887, 505.

[2] Hesse, *Ann. Chem.* (Liebig), 176, 122.

[3] Arndtsen, *Ann. Chim. et Phys.* [3], 54, 403; Přibram, *l.c.*

[4] Schneider, *Ann. Chem.* 207, 257.

ture acting in the same direction, as in general both have the same effect on dissociation. For sugar[1] and the tartrates[2] the alteration with the temperature is scarcely perceptible.

With tartaric acid,[3] warming, like dilution, effects a rise:

Temp.	40 per cent.	20 per cent.	10 per cent.
0°	$a_D = 5{\cdot}53$	$a_D = 8{\cdot}66$	$a_D = 9{\cdot}95$
100°	,, $=17{\cdot}66$	,, $=21{\cdot}48$	,, $=23{\cdot}97$

In the case of malic acid, Pasteur found in the dilute left-handed solution an increase of rotation to the left on warming, which is the result Schneider obtained by dilution. With mandelic acid Lewkowitsch[4] observed a decrease in the rotation on diluting and on warming; with rhamnose Tollens found the same thing.

2. The change of rotation with the concentration is parallel with that effected by the solvent, so that the rotations in other solvents approximate to those in concentrated aqueous solution. Tartaric acid, which in water gradually rotates less to the right as the concentration increases, exhibits in other solvents now a weak right-handed rotation, now even left-handed rotation, as in alcohol.[5]

3. The change of rotation on dilution is in the direction of the numbers obtained for the (acid) salt, and appears to be limited by these numbers. It is, again, in the case of tartaric acid that the subject has

[1] Tuchschmid, *J. prakt. Chem.* [2], **2**, 235.

[2] Krecke, *Arch. Neërl.* **7**, 97.

[3] *Ber.* **16**, 1567.

[4] *Ann. Chem.* (Liebig), **271**, 64.

[5] Přibram, *Wien. Acad.* **97**, 460.

been most thoroughly investigated. The gradual increase of $[\alpha]_D^{25}$ from 8·5° to 14·3° between 50 and 5 per cent. is evidently in the direction of the value found for the acid salt, 29° ; Přibram, indeed, obtained for 0·35 per cent. $[\alpha]_D^{20}=16·3°$, and Krecke at 100° and 10 per cent. observed 23·97°.

Malic acid, right-handed in the concentrated solutions (70 per cent. $[\alpha]_D=+3·34$) and left-handed in dilute solutions (8·4 per cent. $[\alpha]_D=-2·3$), also shows an approximation to its (left-handed) salts; though their (extreme) value ($[\alpha]_D=-9$) is not attained.

Lactic acid, the right rotation of which is diminished by dilution (21·24 per cent. $[\alpha]_D=2·66$; 15·75 per cent. $[\alpha]_D=2·06$), possesses accordingly left rotation in its salts.

4. The acids which undergo no change of rotation on dilution are also those which rotate as strongly as their acid salts. Methoxy- and ethoxy-succinic acids[1] exhibit rotations which scarcely alter with the concentration :

Methoxy-acid	11 per cent.	$[\alpha]_D$ = 33·3°	5·6 per cent.	$[\alpha]_D$ = 33°
Ethoxy-acid	11 „	„ = 33°	5·6 „	„ = 32·5°

These numbers are almost the same as those obtained for the acid salts, viz. $[\alpha]_D=29°$ and 37° respectively.

For quinic acid,[2] also, the rotation is the same, from 2 to 53 per cent. $[\alpha]_D=-43·9°$, while for the salts it is $-49°$.

[1] *Chem. Soc. J. Trans.* 1893, 217, 229.

[2] Hesse, *Ann. Chem.* (Liebig), 176, 124.

The rotation of shikimic acid[1] also alters but little (36·26 per cent. $[\alpha]_D = -204°$; 4·03 per cent. $[\alpha]_D = -183·8°$), while for the ammonium salt it is $-189°$.

The hypothesis of electrolytic dissociation explains these facts to this extent, that it demands that dilution of an acid and salt formation shall bring about equal activity, since both cause the formation of the same ion. For dibasic acids the same holds for the acid salts, because dilution of these acids first liberates a single hydrogen atom.

Evidently, however, there is something else concerned besides electrolytic dissociation, and that is the point of attack offered to the carboxyl group in another part of the molecule, as appears from the following.

5. Great change of rotation on dilution manifests itself specially with the oxy-acids. Malic acid is remarkable in this respect. The change of rotation which we have observed to characterise this acid is no longer found in methoxysuccinic acid and in the corresponding ethyl derivative, nor in chlorosuccinic[2] and acetylmalic[3] acids.

$CO_2H\mathit{C}HOHCH_2CO_2H$	70 % $[\alpha]_D$ =	+ 3·34°	8·4 % $[\alpha]_D$ =	− 2·3°
$CO_2H\mathit{C}HOCH_3CH_2CO_2H$	11 ,, ,, =	33·3°	5·6 ,, ,,	33°
$CO_2H\mathit{C}HOC_2H_5CH_2CO_2H$	11 ,, ,, =	33°	5·6 ,, ,,	32·5°
$CO_2H\mathit{C}HClCH_2CO_2H$	16 ,, ,, =	+ 20·6°	3·2 ,, ,,	+ 21·3°
$CO_2H\mathit{C}HOC_2H_3OCH_2CO_2H$	16 ,, ,, =	− 11°	3·2 ,, ,,	− 10°

Thus when the hydroxyl group disappears the

[1] Eykman, *Ber.* **24**, 1280, 1297. [2] *Ber.* **26**, 215.

[3] Guye, *Arch. Sc. phys. nat.* [3], **29**, 430; Colson, *Compt. Rend.* **116**, 818.

rotation becomes more constant. The peculiar part played by this group is, however, still more plainly manifested in the gradual change which often occurs in oxy-acids after a change of concentration or of temperature. This was first observed in the case of lactic acid,[1] the rotation of which decreased on simple standing of the freshly prepared solution ; it was recently proved in the case of glyceric acid,[2] and is due to etherification or lactone formation, as Wislicenus showed. This will be considered in the next section.

In the oxy-acids, then, the alteration of the rotation on dilution may be due to a phenomenon akin to lactone formation, which also is probably influenced by electrolytic dissociation. Finally, several acids, and not oxy-acids only, possess a double molecule,[3] and accordingly on changing the concentration they may break up in a way which will affect the optical examination. Comparable results for acids are therefore scarcely to be obtained except by an investigation of dilute solutions of the alkali salts.

IV. Influence of Ring Formation on Rotation

The interaction of several of the groups attached to the asymmetric carbon atom, which may be accompanied by ring formation, appears to have a quite extraordinary influence on the magnitude and the sign of the rotation. In the phenomena mentioned above

[1] Wislicenus, *Ann.* **167**, 302.
[2] *Chem. Soc. J. Trans.* 1893, 296.
[3] Bineau, Ramsay, *ibid.* 1893, 1098.

we have already had indications of this, and below the fundamental facts are given.

Lactone formation.—The change of rotation was first observed in the case of lactic acid,

$$CH_3\mathit{C}HOHCO_2H \quad [a]_D = +2^\circ \text{ and } +3^\circ,$$

while the lactone $CH_3\mathit{C}H—CO$ (lactid) has the
\O/

enormous rotation $[a]_D = -86^\circ$.[1] The same change has been observed for glyceric acid,[2] and in the sugar group has indeed become a simple test to distinguish between the isomeric saccharic acids,[3] *e.g.*, of which one forms a lactone, a second a double lactone, a third no lactone. The following table illustrates this :

Lactone formation	$[a]_D$ of the acid	$[a]_D$ of the lactone
Arabonic acid, $CO_2H(\mathit{C}HOH)_3CH_2OH$	$< -8{\cdot}5$ [4]	$-73{\cdot}9^\circ$ [5]
Ribonic acid, ,,	Cd salt $+0{\cdot}6^\circ$ [5]	-18° [5]
Xylonic acid, ,,	-7°	$+21^\circ$ [6]
Gluconic acid, $CO_2H(\mathit{C}HOH)_4CH_2OH$	$-1{\cdot}74^\circ$ [6]	$+68{\cdot}2^\circ$ [7]
Galactonic acid, ,,	$< -10{\cdot}56^\circ$ [6]	$-70{\cdot}7^\circ$ [8]
Mannonic acid, ,,	weak [7]	$+53{\cdot}8^\circ$ [9]
Saccharinic acid, $C_6H_{12}O_6$.	Na salt $-17{\cdot}2^\circ$	$+93{\cdot}6^\circ$ [10]
Isosaccharinic acid, ,, .	left-handed	$+62^\circ$ [10]
Rhamnonic acid, ,, .	$-7{\cdot}67^\circ$	$-38{\cdot}7^\circ$ [8]
Talomucic acid, $CO_2H(\mathit{C}HOH)_4CO_2H$	$> +24^\circ$	$< 7^\circ$ [11]
Saccharic acid, ,,	$+8^\circ$	$+38^\circ$ [12]
Mannosaccharic acid, ,,	weak [5]	$+201{\cdot}8^\circ$ (Double lactone) [13]

[1] Wislicenus, *Ann.* **167**, 302. [2] *Chem. Soc. J. Trans.* 1893, 296.
[3] *Ber.* **23**, 2614. [4] *Ann. Chem.* (Liebig), **260**, 313.
[5] *Ber.* **24**, 4217–4219. [6] *Ann. Chem.* (Liebig), **271**, 78–85.
[7] *Ber.* **23**, 2626. [8] *Ibid.* **23**, 2992.
[9] *Ibid.* **22**, 3218. [10] Tollens, *Kohlehydrate*, 293–295.
[11] *Ber.* **24**, 3628. [12] Tollens, *Kohlehydrate*, 309. [13] *Ber.* **24**, 541.

Where the figures, especially those for the acid, are uncertain, because they are strongly influenced by the time and probably also by the concentration, we cannot avoid the conclusion that lactone formation exerts an influence equally profound; for lactic acid the difference amounts to about 90°, for arabonic acid to 70° or more, the same for gluconic acid, for saccharinic acid 100°, and for the double lactone 200°. If the acids had been investigated as sodium salts, and the lactones pure, some relation would perhaps have been found.[1]

Multi-rotation.—The phenomenon at first known as bi-rotation—where immediately after solution a rotation is observed, which for glucose is twice as large as afterwards—has been shown by further in-

[1] As the result of an investigation made in accordance with this suggestion, the following table has been published. Here the 'molecular rotation' is the specific rotation multiplied by the molecular weight and divided by 1000.

Acid	Molecular rotation		Difference
	Ion	Lactone	
Ribonic . .	+ 0·2	− 3·0	3·2
Gluconic (d) .	+ 1·3 to + 1·8	+ 11 to 12·1	10·8 to 9·2
Mannonic (d and l)	+ 2	− 9·5 to − 9·8	11·7
Saccharinic . .	− 1·1	+ 15·3 to + 15·1	16·3
Isosaccharinic .	− 1·1	+ 10·2	11·3
Saccharic (d) .	− 2·6	+ 7·3 to + 8·0	9·9 to 10·6
Mannosaccharic .	+ 0·2	+ 35·1 to + 35·6 (Double lactone)	35·2
α-Rhamnohexonic	+ 1·3	+ 16·1 to + 16·5	15·0
α-Glucoheptonic .	+ 1·6	− 10·9 to − 11·5	12·8
Gulonic (d and l).	± 2·7	± 9·9	12·6

See W. Alberda van Ekenstein, W. P. Jorissen, and L. Th. Reicher, *Zeitschr. physik. Chem.* **21**, 383.

vestigations, especially those of Tollens,[1] to be a change of rotation which only in the case of glucose amounts to a decrease of about one-half; in other cases there is, indeed, an increase.

Rotation	Initial	Final
Dextrose, $CH_2OH(CHOH)_4COH$ . .	105·2	52·6
Galactose, " . .	117·5	80·3
Levulose, $CH_2OHCO(CHOH)_3CH_2OH$.	− 104	− 92·1 (− 53 at 90°)
Lactose, $C_{12}H_{22}O_{11}$	82·9	52·5
Maltose, "	118·8	136·8
Arabinose, $CH_2OH(CHOH)_3COH$ . .	156·7	104·6
Xylose, " . .	78·6	19·2
Rhamnose, $C_6H_{12}O_6$	− 3·1	+ 8·6
Saccharin, "	+ 92·7	+ 87·5

The phenomenon of multi-rotation corresponds completely to that observed in the case of the lactone-forming acids; if these (galactonic acid, *e.g.*) are set free from their salts in solution, the gradual change of rotation manifests itself here also,[2] only it proceeds faster in the case of the acids. Moreover, the lactone-forming bodies and those possessing multi-rotation are most intimately related to one another; the aldehydes exhibiting multi-rotation—glucose, galactose, arabinose, xylose, rhamnose—correspond to the lactone-forming acids, gluconic and saccharic, galactonic, arabonic, xylonic, and rhamnonic acids.

Then the multi-rotating compounds and the oxy-acids have the hydroxyl and carboxyl groups in common. Finally, since the lactone formation, which is accompanied by the closing of a ring, in general

[1] *Ann.*, **257**, 160; **271**, 61.

[2] Tollens, *Ber.* **23**, 2991.

brings about an increase of rotation, and in the cases now under consideration (maltose excepted) there is a decrease, there is perhaps here a ring opened up. Thus xylose may have been at first

$$CH_2OH\underbrace{CH(CHOH)_2C}_{O}(OH)H,$$

and later, $HOCH_2(CHOH)_3C(OH)_2H$, corresponding to $CH_2OH(CHOH)_3COH$.[1]

And it may be observed that the marked changes of rotation with the concentration and temperature, observed with glucose, galactose, and rhamnose,[2] and especially with levulose and the lactone-forming acids, are to be attributed to changes of equilibrium.

Other internal anhydrides.—There are other isolated cases of great change of rotation through ring formation which are also related to lactone formation.

Propyleneglycol (−4° 55′ 22 mill.) changes the sign of the rotation on being transformed into propyleneoxide (+1° 10′ 22 mill.).[3] The same is the case with left diacetyltartaric acid, $a_D = -19{\cdot}23$, which forms a right-handed anhydride, $a_D = +62{\cdot}04$.[4] Finally, phenylbromolactic acid yields a much stronger phenoxacrylic acid of reverse rotation.[5]

[1] In the case of glucose, according to Trey, hydration does not take place (*Zeitschr. physik. Chem.* **18**, 193).

[2] Tollens, *Kohlehydrate*, and *Ann.* **271**, 61.

[3] *Jahresber.* 1881, 513.

[4] *Ibid.* 1882, 856. [This change of sign does not occur with acetylmalic acid. (*Ber.* **26**, R. 371, 492.)]

[5] *Ber.* **24**, 2830.

The dibromoshikimic acid, $C_7H_{10}Br_2O_5$ $a_D = -58$, gives a right-handed bromo-lactone,

$$C_7H_9BrO_5(+22°).$$[1]

Boric acid and polyatomic alcohols.—Now that the increase of rotation through ring formation has been established, the very considerable rise of rotation observed on addition of boric acid is seen in another light. Such is the effect of this addition that, as is well known, it was only by this means that activity could be demonstrated in the case of mannite, sorbite, arabite, &c. If we consider now the more recent observations,[2] especially those of Magnanini, we see in the first place that the proved diminution of the number of molecules involves the hypothesis that an addition product is formed. In the next place, in view of the fact that only polyatomic alcohols (including erythrite)[3] and oxy-acids are affected by boric acid, while mannite with six hydroxyl groups demands three molecules of boric acid, there must be two hydroxyl groups connected with one boric acid molecule, and we come of necessity to the hypothesis that the following ring is formed:

```
C—O\
|    >B—O—H,
C—O/
```

[1] Eykman, *Ber.* **24**, 1293. See also the high rotation of methyl-glucoside, &c. (Fischer, *Ber.* **26**, 2400).

[2] *Zeitschr. physik. Chem.* **6**, 58; *Gazz. Chim.* **11**, 8, 9; 1891. (Ref. *Zeitschr. physik. Chem.* **9**, 230.)

[3] Klein, *Compt. Rend.* **86**, 826; **99**, 144.

which is in harmony with the other properties (acid character, depression, conductivity).

Tartar emetic and analogous substances.—The enormous increase of rotation which tartaric acid in its salts ($[a]_D$=20 to 30) undergoes on transformation into tartar emetic, and the analogous phenomenon in the case of malic acid (salts, $[a]_D$= −10 to 20; antimony derivative, + 115[1]) suggest similar considerations. The fact that only oxy-acids yield compounds of this kind, the formula of tartar emetic $(C_4H_4O_6K)_2Sb_2O_2.H_2O$,[2] the anomalous reactions, the depression ($i=1\frac{1}{2}$)[3] are in the most perfect harmony with the following hypothesis as to the constitution:

```
CO2K                          CO2K
HCOH                          CHOH
HC—O                        O—CH       + 2H2O.
 |    >Sb—O—Sb<               |
OC—O                        O—CO
```

Salts of polyvalent metals and polybasic acids.—The change of rotation in the salts of polyvalent metals, to which we have already called attention, is partly due to the fact that in general they do not undergo electrolytic dissociation to the same extent as the alkali salts.

But this change of rotation is especially noticeable when the acid is polybasic, so that here too interaction of the two carboxyl groups (ring formation) is possible. The observed alterations in such

[1] Landolt, *Opt. Dreh.-vermögen.* 221. [2] *Ber.* **16**, 2386.
[3] *Zeitschr. physik. Chem.* **9**, 484.

cases are therefore collected here. First we have Schneider's values for a_D for malates :

	Ba	K_2	Na_2	Li_2	$(NH_4)_2$
20 per cent. >	+15	− 9·6	− 8·2	− 8·6	− 8·7
— „ <	− 5	−11·5	−13·1	−13·9	−11·2

Then the striking results with methoxy- and ethoxy-succinic acid :

Salts	Methoxy-acid		Ethoxy-acid	
$(NH_4)_2$ . .	5·76 per cent.	15·2	5·22 per cent.	22·2
„ . .	2·82 „	15·1	1·48 „	22·8
K_2 . . .	12·16 „	14·3	—	—
„ . . .	5·02 „	14·2	—	—
Ca . .	5·31 „	− 12·7	3·04 per cent.	+ 10·4
„ . . .	2·21 „	+ 5·2	1·79 „	+ 14·1
Ba . .	26·12 „	− 27·3	25·08 „	− 8
„ . . .	1·15 „	+ 6·1	4·56 „	+ 11·7

V. Rotation of Non-electrolytes. Hypotheses of Guye and Crum Brown

In the case of non-electrolytes, the most important fact concerning the material that has been collected is that as a rule all exact comparison of results is impossible. These compounds have not yet been examined in the same solvent, diluted, and with due regard to the molecular weight and to the possible action of the solvent.

That a determination of molecular weight must be made follows from the observation of Haller,[1] who finds for left isocamphol (borneol) in alcohol, $a_D = 33°$; in benzene and its homologues, $a_D = 19°$. According to Paterno, hydroxyl derivatives possess in benzene a

[1] *Compt. Rend.* **112**, 143.

double molecule. The known borneol according to Beckmann,[1] does not do this, and has also the same rotation in benzene as in alcohol, $a_D = 37°$. And Freundler[2] has recently shown that in the case of ethereal tartrates the change of rotation by the solvent is accompanied by a change of molecular weight.

The views of Guye and of Crum Brown deserve especial notice. The latter[3] proposes to establish by experiment a function, K (which perhaps alters for the temperature, &c.), for each of the groups attached to the asymmetric carbon; the rotation would be determined by the difference of these functions. From the material at hand he thinks it may be concluded that the function for any group rises as the group increases.

The objection to this hypothesis is, as the author himself observes, that in it the mutual action of the groups plays no part. But in view of what has just been said about the influence of the solvent, and of ring formation on the rotation, this mutual action must be essential. Guye[4] starts on a broader basis, viz. the whole configuration of the molecule, and he proceeds to determine numerically its degree of disymmetry, by the displacement of its centre of

[1] *Zeitschr. physik. Chem.* **6**, 440. In accordance with this view the hydroxyl-free phenylurethane, obtained from isocamphole, shows no change of rotation.

[2] *Compt. Rend.* **117**, 556.

[3] *Proc. Roy. Soc. Edinb.* June 1890.

[4] *Compt. Rend.* March 1890; *Ann. Chim. phys.* [6], **25**, 145; *Arch. Sc. phys. Nat.* [3], **26**, 97, 201, 333; *Rev. scientifique*, **49**, 265.

gravity in relation to the six planes of symmetry of the regular tetrahedron. Six values, $d_1 \ldots . d_6$, are thus obtained, the product of which, called product of asymmetry, determines the rotation :

$$P = d_1\, d_2\, d_3\, d_4\, d_5\, d_6.$$

This product satisfies the main condition, that when only two of the groups are equal, one of the displacements (the one referred to the plane of symmetry between the two groups) becomes *nil*, and consequently P is also *nil*, which corresponds with inactivity.

These values d, however, are difficult to determine ; they are certainly influenced by the weights of the groups and by their distances,[1] and the first thing is to determine the part played by the weight. The displacement is then determined by the difference of weight, and we have as a concrete expression of this :

$$P = (g_1 - g_2)\,(g_1 - g_3)\,(g_1 - g_4)\,(g_2 - g_3)\,(g_2 - g_4)\,(g_3 - g_4),$$

where $g_1 \ldots . g_4$ are the group-weights in question.

This expression is not a necessary consequence of Guye's conception, but only a formulation of it upon certain assumptions made for the sake of simplicity. It is to be regarded as a special case of the view of Crum Brown, according to which K and g are identical. Finally, we may repeat that the

[1] Compare Frankland and Wharton (*J. Chem. Soc.* 1896, 1309) on the methyl and ethyl esters of o-, m-, and p-ditoluyltartaric acid ; also Guye, *Bull. Soc. Chim. Paris*, [3], 15, 1137.

essential requisite, that $P=o$ when two groups are identical, is fulfilled ; and that if two groups, g_3 and g_4, *e.g.*, change places, the sign of P is simply reversed, its numerical value remaining the same.

From this view the following novel and essential consequences result. If the groups are in the following order :

$$g_4 > g_3 > g_2 > g_1,$$

and the substance is, say, right-handed, then when g_4 is replaced by smaller and smaller groups, we may expect :

1. Diminution of the right rotation for $g_4 > g_3$;
2. Inactivity when $g_4 = g_3$;
3. Left rotation, increasing to a maximum and then diminishing, when $g_3 > g_4 > g_2$;
4. Inactivity when $g_4 = g_2$;
5. Right rotation, increasing to a maximum, and then diminishing, when $g_2 > g_4 > g_1$;
6. Inactivity when $g_4 = g_1$;
7. Left rotation, increasing, when $g_4 \quad g_1$.

Thus, when one of the groups gradually passes from the maximum to the minimum the sign of the rotation will change four times.

Let us consider first the derivatives of active amylalcohol, $C_2H_5(=29)CH_3.CH.CH_2OH$. The substances are arranged in the order of the magnitude of the radical replacing CH_2OH, and it is seen that, in general, increase of the largest group leaves the sign of the rotation unaltered :

1. Aldehyde, $COH=29$ $a_D=+0° 42'$ (10 Dec.)
2. Amine, $CH_2NH_2=30$ $a_D=-3° 30'$ (10 Dec.)
3. Alcohol, $CH_2OH=31$ $[a]_D=-5° 2'$.
4. Nitrile, $CH_2CN=40$ $a_D=+1° 16'$ (10 Dec.)
5. About sixty compounds between 4 and 6, all right-handed.
6. Iodide, $CH_2I=141$ $a_D=+8° 20'$ (10 Dec.)

The change of sign observed in the case of the amine and the alcohol should, however, not occur till below 29°.

It follows that change of sign can be brought about by causes other than change of weight. In this connection the cases where, as in amylaldehyde, there are two groups of equal weight are especially convincing. Here we do not find inactivity, which Guye's formula would demand. Such cases are:

Dimethylic diacetyl tartrate,

$$CO_2CH_3(CHOC_2H_3O)_2CO_2CH_3 :$$
$$CO_2CH_3=OC_2H_3O=59 \text{ (left-handed).}$$

Diethylic dipropionyl tartrate,

$$CO_2C_2H_5=OC_3H_5O=73 \text{ (slightly right-handed).}$$

Dipropylic dibutyryl tartrate,

$$CO_2C_3H_7=OC_4H_7O=87 \text{ (right-handed).}$$

Acetylmalic acid, $CO_2HCHOC_2H_3OCH_2CO_2H$:

$$OC_2H_3O=CH_2CO_2H=59 \text{ (left-handed).}$$

Ethylmalic acid, $CO_2HCHOC_2H_5CH_2CO_2H$:

$$CO_2H=OC_2H_5=45 \text{ (right-handed).}$$

The following table of such esters of tartaric,[1]

[1] Pictet, *Arch. des Sc. phys. et nat.* [3], **7**, 82; Freundler, *Compt. Rend.* **115**, 509.

glyceric,[1] and valerianic[2] acids as have been investigated yields the same result, namely, that the weight of the groups acts in the sense demanded by Guye's fundamental conception, but not strictly according to the formula chosen by him as a first approximation.[3]

	CH_3	C_2H_5	C_3H_7	C_3H_7 iso	C_4H_9	C_4H_9 iso	C_7H_7
Tartaric acid . .	2·1	7·7	12·4	14·9	15·9	—	—
Acetyltartaric acid .	−14·3	5	13·5	—	17·8	11·3	—
Propionyltartaric acid	−12	0·3	7·9	—	—	9·2	—
Butyryltartaric acid .	−13	−1	5·4	—	—	7·1	—
Benzoyltartaric acid .	−88·8	−60	{>−60 <−42}	—	—	−42	—
Glyceric acid . .	−6·8	−9·2	−12·9	−11·8	−11	−14·2	—
Valerianic acid . .	16·8	13·4	11·7	—	10·6	10·5	5·31

In the glyceric acid derivatives, $CH_2OH\mathit{C}HOHCO_2X$, the rotation is seen to rise as the largest group, CO_2X, becomes larger.

The case of tartaric acid is somewhat more complicated. In the first place there are two asymmetric carbon atoms, but these being perfectly identical we may confine ourselves to the consideration of one. But, further, in the derivatives there are always two groups which alter. If we set out the groups thus:

$$H=1 < OH=17 < CO_2H=45 < CHOHCO_2H=75,$$

we see that, in the esters in which carboxyl hydro-

[1] Frankland and MacGregor, *Chem. Soc. J. Trans.* 1893, 524.

[2] Guye, *Compt. Rend.* **116**, 1454. Is the decrease of rotation with the group-weight due to the increased formation of the racemoïd form on distillation, caused by the higher boiling-point?

[3] In confirmation of this see Walden, *Zeitschr. physik. Chem.* **15**, 638; I. Welt, *Compt. Rend.* **119**, 885; *Ann. Chim. Phys.* [7], **6**, 115; Ph. A. Guye and L. Chavanne, *Compt. Rend.* **119**, 906; **120**, 452. Compare J. W. Walker, *J. Chem. Soc.* **67**, 914, and Purdie and Williamson, *l.c.* p. 957.

gen is replaced, the two largest groups increase, and therefore the rotation; in those in which hydroxyl hydrogen undergoes substitution, OH=17 increases, and also the largest group: a change of sign is therefore to be expected, and at the same time an increase in the numerical value. Both occur; only the change of sign does not exactly correspond with the equality of the group-weights.

Further, we must emphasise the fact that, in isomeric compounds, groups of equal weight do not correspond to equal rotations. Among glyceric esters, propyl- and iso-propyl, butyl- and iso-butyl have not the same action; with tartaric acid the case is the same, but not with valerianic acid. But whether in the first two cases the difference is as great as the figures indicate is uncertain, as it is doubtful how far they can be compared. Thus Freundler found that ethylic diacetyl tartrate rotates, in alcohol, +1·02 instead of +5.

Finally, it is a striking fact, in agreement with Guye's conception, that the very high rotations are observed among compounds of high molecular weight. One example of this is seen in methylic benzoyl tartrate, −88·8°. Then we have the small rotation of + 2° for lactic acid, as compared with −21° for oxybutyric, and −11° for leucic acid, 71° for tropaic acid, −156° for mandelic acid, and −135° for isopropylphenylglycollic acid. Perhaps in the last the effect of ring formation is superadded. It is a fact that the highest known rotations are found among the alkaloids and santonine derivatives (over

300° for quinidine, 700° for santonine), where several rings and high molecular weight coexist. Of course, the converse of this rule does not hold. Even when the molecular weight is high, identity among the groups annihilates the rotation, and similarity among the groups perhaps reduces it to small proportions.

VI. More Complicated Cases

Several asymmetric groups in one molecule.—So far we have dealt chiefly with the simplest cases, with a single asymmetric carbon atom. It remains to add a few words on more complicated compounds, which may throw some light on the subject. In the first place we may consider the idea expressed in my former pamphlet[1] that when there are several asymmetric carbon atoms their action is to be added or subtracted. Thus for the four pentose types, $COH(CHOH)_3CH_2OH$, we should have the following rotations :

No. 1	No. 2	No. 3	No. 4
$+A$	$+A$	$+A$	$-A$
$+B$	$+B$	$-B$	$+B$
$+C$	$-C$	$+C$	$+C$

and since the sum of No. 2, No. 3, and No. 4 is equal to $A+B+C$, the rotation of arabinose (probably the highest) should be equal to the rotations of xylose, ribose, and the expected fourth type[2] taken together.

For the asymmetric compounds of the saccharic acid group a similar conclusion may be drawn. The four active types would have the following rotations :

[1] See Preface.

[2] Discovered since, and called lyxose.

No. 1	No. 2	No. 3	No. 4
$+A$	$+A$	$+A$	$+A$
$+B$	$+B$	$+B$	$-B$
$+B$	$+B$	$-B$	$-B$
$+A$	$-A$	$+A$	$+A$
$2(A+B)$	$2B$	$2A$	$2(A-B)$
CO_2H	CO_2H	CO_2H	CO_2H
HCOH	HCOH	HCOH	HCOH
HOCH	HOCH	HOCH	HCOH
HCOH	HCOH	HOCH	HOCH
HOCH	HCOH	HOCH	HOCH
CO_2H	CO_2H	CO_2H	CO_2H
Idosaccharic acid	Saccharic acid 8°	Talomucic acid 29°	Manno-saccharic acid; weak.

The large rotation $2(A+B)$ might belong to the first type and would amount to 37°. This corresponds to the constitution in that neither the inner nor the outer asymmetric carbon atoms are symmetrically opposed. Then saccharic acid corresponds to $2B$, because in its configuration the two outer carbon atoms are symmetrically opposed; for similar reasons talomucic acid corresponds to $2A$. For mannosaccharic acid we should then have about 20° (29° − 8°); all that is known is that it possesses slight activity. Since the acids readily form lactones an exact investigation of the sodium salts in not too concentrated solution seems to be the only way to arrive at definite results.

Further, it is to be noted that the outer asymmetric carbon atoms cause a rotation of 29°, the inner a rotation of 8°, and this greater influence of the excentric carbons accords with Guye's theories.

Influence of the type.—In the second place we must note the fact that the magnitude of the rotation is to a certain extent determined by the type of the compound.

We have already observed (p. 147) what an effect lactone formation has on the rotation, an effect which often amounts to about 80°; and how ring formation in other cases causes a fairly definite change of rotation. We saw, further, that in many cases stereomers though not enantiomorphous possess equal rotation (p. 74). Now it has been observed that in chemically related compounds there are often found rotations of a similar order of magnitude.

1. Thus, the alcohols of the type

$$CH_2OH(CHOH)_nCH_2OH$$

have a remarkably small rotation, often noticeable only after addition of borax:

Arabite . . .	$CH_2OH(CHOH)_3CH_2OH$	– 5° in borax
Mannite . . .	$CH_2OH(CHOH)_4CH_2OH$	almost *nil*
Sorbite . . .	,,	1°
Perseite [1] . . .	$CH_2OH(CHOH)_5CH_2OH$	8° in borax
α-Glucosectite [2] .	$CH_2OH(CHOH)_6CH_2OH$	2°

It is very remarkable that in the hexatomic alcohol inosite, $C_6H_6(OH)_6$—which in composition resembles mannite, but as a hexamethylene derivative belongs to another type—we observe at once a comparatively strong rotation of 65° (caused by the ring formation).

[1] *Ber.* **23**, 2226. [2] *Ann.* **270**, 64.

2. The amido-acids exhibit rather low rotations :

Leucine . .	$C_4H_9\mathit{C}HNH_2CO_2H$	14° HCl; 6° NH_3
Phthalyl derivative . .	$C_4H_9\mathit{C}HN(C_6H_4C_2O_2)CO_2H$	− 22° C_2H_6O
Cystine . .	$CH_3\mathit{C}(NH_2)(SH)CO_2H$	− 8° H_2O
Phenylcystine .	$CH_3\mathit{C}S(C_6H_5)NH_2CO_2H$	< − 4° NaOH
Bromine derivative . .	$CH_3\mathit{C}S(C_6H_4Br)NH_2CO_2H$	− 4° NaOH
Acetyl derivative	$CH_3\mathit{C}S(C_6H_5)NHAcCO_2H$	− ? C_2H_6O; 5° NaOH
Bromacetyl derivative . .	$CH_3\mathit{C}S(C_6H_4Br)NHAcCO_2H$	− 7° C_2H_6O; 8° NaOH
Tyrosine . .	$C_6H_4OHCH_2\mathit{C}HNH_2CO_2H$	− 8° HCl; − 9° KOH
Phenylamidopropionic acid	$C_6H_5CH_2\mathit{C}HNH_2CO_2H$	− 35° H_2O
Asparagine .	$CO_2H\mathit{C}HNH_2CH_2CONH_2$	− 8° H_2O; + 37° HCl
Aspartic acid .	$CO_2H\mathit{C}HNH_2CH_2CO_2H$	− 4° H_2O; + 25° HCl
Glutamic acid .	$CO_2H\mathit{C}HNH_2C_2H_4CO_2H$	10° H_2O; 26° HCl; − 5° CaO_2H_2
Glutamine .	$CO_2H(C_3H_5NH_2)CONH_2$	slightly right-handed, H_2SO_4
Chitamic acid .	$C_6H_{13}NO_6$	+ 1·5° H_2O

The weak activity of the amido-acids is probably the reason why no rotation has as yet been discovered in the case of serine, alanine, &c.

3. Among the lactones of the sugar group, &c., larger variations occur; the values, however, do not exceed 90°, which amount is attained by the simplest, lactid. This is probably due to the fact that the oxy-acids have generally a low rotation, and that between acid and lactone there is usually a difference of 80°. The lactones of the following oxy-acids may be cited :

Lactic acid	$CH_3\mathit{C}HOHCO_2H$	− 86°
Arabonic acid . . .	$CH_2OH(\mathit{C}HOH)_3CO_2H$	− 74°
Ribonic acid	,,	− 18°
Xylonic acid	,,	+ 21°

Saccharinic acid .	. $C_6H_{12}O_6$	+ 94°
Isosaccharinic acid	. „	+ 62°
Rhamnonic acid .	. „	− 38°
Gluconic acid .	. $CH_2OH(CHOH)_4CO_2H$	+ 68°
Galactonic acid .	. „	− 71°
Mannonic acid .	. „	+ 54°
Talomucic acid .	. $CO_2H(CHOH)_4CO_2H$	7°
Saccharic acid .	. „	38°
Mannoheptonic acid	. $CH_2OH(CHOH)_5CO_2H$	− 74°
α-Glucoheptonic acid	. „	− 68°
β-Glucoheptonic acid	. „	+ 23°
Gluco-octonic acid	. $CH_2OH(CHOH)_6CO_2H$	+ 46°
Mannononic acid .	. $CH_2OH(CHOH)_7CO_2H$	− 41°

For the double lactone of mannosaccharic acid the more than double value of 202° is attained.

The small rotations of the oxy-acids are shown in the following table :

Malic acid . .	. $CO_2HCHOHCH_2CO_2H$	weak + or −
Tartaric acid .	. $CO_2H(CHOH)_2CO_2H$	„ „
Oxyglutaric acid .	. $CO_2HCHOHC_2H_4CO_2H$	− 2°
Trioxyglutaric acid	. $CO_2H(CHOH)_3CO_2H$	− 23°
Arabonic acid .	. $CH_2OH(CHOH)_3CO_2H$	less than − 8°
Ribonic acid .	. „	Cd salt + 1°
Xylonic acid .	. „	− 7°
Isosaccharinic acid	. $C_6H_{14}O_7$	Na salt − 17°
Rhamnonic acid .	. „	− 8°
Gluconic acid .	. $CH_2OH(CHOH)_4CO_2H$	− 2°
Galactonic acid .	. „	less than − 11°
Mannonic acid .	. „	weak
Talomucic acid .	. $CO_2H(CHOH)_4CO_2H$	29°
Saccharic acid .	. „	8°
Mannosaccharic acid .	„	weak

For the acids of the type $CH_2OH(CHOH)_4CO_2H$, beginning with gluconic acid, the rotations seem to be extremely small. It is a striking fact that if a benzene nucleus (ring formation) is introduced into these oxy-acids relatively high values result:

Lactic acid	$CH_3\mathit{C}HOHCO_2H$	+ 2°
Oxybutyric acid . . .	$CH_3\mathit{C}HOHCH_2CO_2H$	− 21°
Leucic acid	$C_4H_9\mathit{C}HOHCO_2H$	− 4°
Mandelic acid . . .	$C_6H_5\mathit{C}HOHCO_2H$	± 156°
Tropaïc acid . . .	$C_6H_5\mathit{C}H(CH_2OH)CO_2H$	+ 71°
Propylmandelic acid . .	$C_3H_7C_6H_4\mathit{C}HOHCO_2H$	± 135°

4. Among the aldehyde sugars, pentoses, glucoses, heptoses, &c., a difference amounting frequently to 50° is caused by multi-rotation, and the maximum value, somewhat above 150°, is thus attained; otherwise the values are below 100°, as with the lactones:

Arabinose	$COH(\mathit{C}HOH)_3CH_2OH$	105° (158°)
Xylose	„	19° (79°)
Dextrose	$COH(\mathit{C}HOH)_4CH_2OH$	53° (105°)
Galactose	„	80° (118°)
α-Glucoheptose . . .	$COH(\mathit{C}HOH)_5CH_2OH$	− 20° (− 25°)
Mannoheptose . . .	„	85°
α-Gluco-octose . . .	$COH(\mathit{C}HOH)_6CH_2OH$	− 50°
Manno-octose . . .	„	− 3°
Mannononose . . .	$COH(\mathit{C}HOH)_7CH_2OH$	50°

5. **Remarkable cases. Cystine derivatives.**—The determinative action of the type is especially striking in the case of cystine, $CH_3\mathit{C}.NH_2.SH.CO_2H$. The small rotation (−8°) characteristic of the amido-acids is maintained when the hydrosulphyl hydrogen is replaced by phenyl, and by bromophenyl (group weight, $C_6H_4Br=156$), also in the corresponding acetyl derivatives (−7°, p. 163). Upon oxidising to cystin,[1] $CH_3\mathit{C}NH_2.CO_2H.SSCO_2H.NH_2.\mathit{C}CH_3$, how-

[1] Baumann, *Zeitschr. physiol. Chem.* **8**, 305; Mauthner, *l.c.* **7**, 222.

ever, we get at once the enormous value $[\alpha]_D = -214°$ (group weight, $S.CO_2H.NH_2.CCH_3 = 120$).

Shikimic acid.[1]—The remarkably high rotation of the derivatives of this acid—

```
          HCOH
         /    \
     H2C       CHOH
       |        |
    HOHC        CH
         \    //
           C
          CO2H
```

appears to be connected with the partial saturation of the benzene ring, as the following table shows :

Without saturation	$[\alpha]_D$	After saturation	$[\alpha]_D$
Acid	− 184°	Dihydrogen product .	− 18°
Ammonium salt . .	− 166°	Dibromine	− 58°
Triacetyl acid . .	191°	Bromolactone . . .	+ 22°
Triacetylethylic ester .	− 174°	Dioxy-acid	− 28°

It is noteworthy that here, too, the lactone formation has the very considerable influence already mentioned (p. 147), and indeed to about the same extent.

Limonenenitroso chloride and derivatives.[2]—The striking fact here is that substitution of amine residues for the chlorine in

```
         ClCC3H7
         /    \
      HC       CH2
      ||        |
      HC        CNOH
         \    /
         HCCH3
```

[1] Eykman, *Ber.* **24**, 1285.
[2] Wallach, *Ann.* **252**, 151.

reduces the remarkably high rotation :

α-Nitrosylchloride	− 315°	α-Nitrolepiperidine	− 68°
		α-Nitrolebenzylamine	− 164°
β-Nitrosylchloride	− 242°	β-Nitrolepiperidine	+ 60°

Nitro-camphor and derivatives.[1]—In connection with the amount of the rotation, the nitro-derivative of camphor, perhaps,

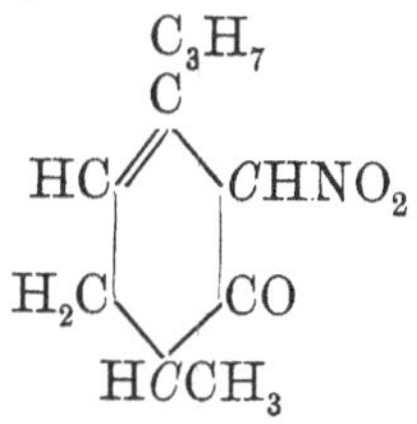

$$\begin{array}{ccc} & C_3H_7 & \\ & C & \\ HC & & CHNO_2 \\ H_2C & & CO \\ & HCCH_3 & \end{array}$$

is highly interesting.

Probably no compound undergoes such a sudden change of rotation with the solvent and the concentration :

$a_j = -\ 140°$ (0·7 per cent. in benzene)
$a_j = -\ 102°$ (5·2 ,, ,, ,,)
$a_j = -\ 7°$ (3 ,, ,, alcohol)

Further, the salts rotate very strongly, and in the opposite direction :

Zinc salt . . . $a_j = +\ 275°$
Sodium salt . . $a_j = +\ 289°$

Santonine derivatives.[2]—Although their constitution has not yet been sufficiently investigated, the

[1] Cazeneuve, *Compt. Rend.* **103**, 275 ; **104**, 1522 ; *Jahresber.* 1888, 1636.

[2] Carnelutti, Nasini, *Ber.* **13**, 2210 ; **22**, Ref. 732 ; **24**, Ref. 909 ; **25**, Ref. 938.

members of this group are remarkable on account of their enormous rotations, which in the case of santonid and parasantonid are as high as $[a]_D=745$ and 892. Lactone formation plays here, as in the whole santonine group, its usual part, raising the rotation, as the following table shows :

Acids	a_D	Lactones	a_D
Santoninic acid, $C_{15}H_{20}O_4$	− 26°	Santonine, $C_{14}H_{18}O_3$	− 174°
Santonic acid, $C_{15}H_{20}O_4$	− 70°	Santonid, $C_{15}H_{18}O_3$	+ 745°
Santononic acid, $C_{30}H_{38}O_6$	+ 37°	Santonone, $C_{30}H_{34}O_4$	+ 129°
Isosantononic acid, ,,	− 40°	Isosantonone, ,,	+ 265°

Finally, it may be observed that the cause of the remarkably high rotation appears to be akin to the cause of the colour of organic compounds.

STEREOCHEMISTRY OF NITROGEN COMPOUNDS

SINCE on the one hand the isomeric benzildioxine discovered by Goldschmidt[1] was proved by Meyer and Auwers[2] to be structurally identical with the one formerly known, and since on the other hand Le Bel[3] obtained active ammonium derivatives, the stereochemistry of nitrogen compounds, which I have already had occasion to deal with,[4] has acquired practical interest.

To begin with that which is simplest, let us in the first place consider the compounds of trivalent nitrogen.

I. TRIVALENT NITROGEN

A. TRIVALENT NITROGEN WITHOUT DOUBLE LINKAGE

Here, where we have to do with the configuration of four atoms or groups, NXYZ, the case is still simpler than with carbon, where there were five, $C(R_1R_2R_3R_4)$, to consider. Putting the matter quite

[1] *Ber.* **16**, 1616, 2176. [2] *Ibid.* **21**, 784.

[3] *Compt. Rend.* **112**, 724.

[4] *Maandblad voor Natuurwetenschappen*, 1877; *Ansichten über org. Chemie*, 80, 1878.

generally—that is, without for the present calling in the aid of the tetrahedron—we may say in the latter case that, given the identity of two groups, *e.g.* R_3 and R_4, a mechanical necessity demands that these two groups shall be similarly situated with regard to the whole, which only happens if they are symmetrically arranged with regard to the plane passing through CR_1R_2. This brings us at once to the tetrahedral arrangement; only it may as well be R_1 or R_2 as the carbon which occupies the centre. The latter is only the case on the assumption of directive forces proceeding from the carbon atom.

In the case of nitrogen derivatives, NXYZ, we should from general mechanical considerations arrive at a tetrahedron of some form, which of course would be unsymmetrical and would lead to optical isomerism. Attempts at 'doubling,' made by Kraft [1] and by Behrend and König with $NH(C_2H_5)=C_7H_7$, p-tolylhydrazine, hydroxylamine bases (NHROH),[2] gave negative results. It is therefore not improbable that the groups NXYZ lie in one plane,[3] which

[1] *Ber.* **23**, 2780.

[2] *Ann.* **263**, 184. Also Ladenburg (*Ber.* **26**, 864) tried in vain to obtain optically active methylaniline, tetrahydroquinoline, and tetrahydropyridine.

[3] Further evidence of this has been supplied by the discovery of two stereomeric compounds of the ammonia type, which proved to be inactive.

Isomers having the plane formulæ,

```
X\   /Y        X\   /Z
   N              N
   |              |
   Z              Y
```

are of course impossible, because these configurations are identical.

again points to the existence of directive forces, in this case proceeding from the nitrogen.

B. TRIVALENT NITROGEN DOUBLY LINKED WITH CARBON

The oximes.—The first remarkable isomerism among nitrogen isomers, which indicated the existence of stereochemical relations, was that of the oximes, which are known to contain the group C=NOH.

It was found to be a perfectly general rule that isomerism occurs when the groups attached to carbon are different, as the following table shows:

Aldoximes	HXCNOH
Ethylaldoxime[1]	. CH_3HCNOH
Propionaldoxime[2] . . .	. C_2H_5HCNOH

But if X, Y, Z are bunched together by their mutual attraction, then

```
X           X
|  /Y        |  /Z
| /    and  | /
N—Z         N—Y
```

represent two different configurations. Accordingly the stereomers in question, which are condensation products of acetaldehyde with asym. m-xylidene, may be represented by the formulæ:

```
CH(CH3).CH2CHO              CH(CH3).CH2CHO
   •                            •
N.C6H3(CH3)2        and      N.H
   •                            •
H                            C6H3(CH3)2
```

(v. Miller and Plöchl, *Ber.* **29**, 1462, 1733). The presence of an asymmetric carbon indicates that each isomer should be divisible into two active forms. It must be noted that the persistence of the isomerism in the compounds of the two substances has not yet been established.

[1] Franchimont, *Versl. Kon. Akad. Amsterdam*, 1892; *Rec. Pays-Bas*, **10**, 236.

[2] Dunstan, *Chem. Soc. J. Proc.* 1893, 76; *Ber.* **26**, 2856.

Aldoximes	HXCNOH
Furfuraldoxime [1]	$C_4H_3OHCNOH$
Thiophenaldoxime [1] . . .	$C_4H_3SHCNOH$
Aldoximeacetic acid [2] . . .	CO_2HCH_2HCNOH
Benzaldoxime [3]	C_6H_5HCNOH
p-, o-, and m-Nitrobenzaldoxime [4]	$C_6H_4(NO_2)HCNOH$
o-, m-, and p-Chlorbenzaldoxime [5]	$C_6H_4ClHCNOH$
3-, 4-Dichlorbenzaldoxime [6] .	$C_6H_3Cl_2HCNOH$
Cuminaldoxime [7]	$C_6H_4(C_3H_7)HCNOH$
Anisaldoxime [8]	$C_6H_4(OCH_3)HCNOH$
Ketoximes	**XYCNOH**
Oximidosuccinic acid [9] . .	$CO_2HCNOHCH_2CO_2H$
Phenylchlorphenyl [10] . . .	$C_6H_5.C_6H_4ClCNOH$
,, bromphenyl [10] . . .	$C_6H_5.C_6H_4BrCNOH$
,, tolyl [11]	$C_6H_5.C_7H_7CNOH$
,, anisyl [8]	$C_6H_5.C_6H_4(OCH_3)CNOH$
,, ethylphenyl . . .	$C_6H_5.C_6H_4C_2H_5CNOH$
,, propylphenyl . . .	$C_6H_5.C_6H_4C_3H_7CNOH$
,, isopropylphenyl . .	,,
,, amidophenyl . . .	$C_6H_5C_6H_4NH_2CNOH$
,, oxyphenyl [12] . . .	$C_6H_5.C_6H_4OHCNOH$
,, xylylphenyl [13] . . .	$C_6H_5.C_8H_9CNOH$
Benzoin [14]	$C_6H_5.CNOHCH_2C_6H_5$
Benzil [15]	$C_6H_5.CNOHCOC_6H_5$

[1] Goldschmidt and Zanoli, *Ber.* **25**, 2573.

[2] Hantzsch, *ibid.* **25**, 1904.

[3] Beckmann, *Ber.* **22**, 429, 514; **23**, 1531, 1588.

[4] Goldschmidt, *ibid.* **23**, 2163; **24**, 2547; Behrend, *l.c.* 3088; Hantzsch, *l.c.* **23**, 2170; Goldschmidt and v. Rietschoten, *l.c.* **26**, 2100.

[5] Behrend and Niessen, *Ann.* **269**, 390; Erdmann and Schwechten, *ibid.* **260**, 60.

[6] *Ibid.* **260**, 63.

[7] Goldschmidt and Behrend, *Ber.* **23**, 2175.

[8] Beckmann, *Ber.* **23**, 1687; *vide* also Goldschmidt, *ibid.* **23**, 2163; Hantzsch, *ibid.* **24**, 36, 3479.

[9] Cramer, *ibid.* **24**, 1198.

[10] Auwers and Meyer, *ibid.* **23**, 2063. [11] Wegerhof, *Ann.* **252**, 11.

[12] Hantzsch, *Ber.* **24**, 5, 3479.

[13] Smith, *ibid.* **24**, 4029. [14] Werner, *ibid.* **23**, 2333.

[15] Auwers and Meyer, *ibid.* **22**, 537; Beckmann, *ibid.* **22**, 514.

Ketoximes	XYCNOH
Carvoxime [1]	$C_9H_{14}CNOH$
Thienylphenyl [2]	$C_4H_3SCNOHC_6H_5$
Acetacetic ester [3]	$CH_3CNOHCO_2C_2H_5$
Papaveraldoximes [4]	$C_6H_3(CH_3O)_2CNOHC_9NH_5(CH_3O)_2$
Phenylketoximepropionic acid [5]	$C_6H_5CNOHCH_2CH_2CO_2H$
Phenylketoximecarboxylic acid [6]	$C_6H_5CNOHCO_2H$
Hydroxamic acid	HOXCNOH
Ethylbenzhydroxamic acid [7]	$C_2H_5O.C_6H_5CNOH$

It must be mentioned that the only possible aldoxime which contains two similar atoms attached to carbon, H_2CNOH, exhibits no isomerism, nor does the diphenyl derivative among the ketoximes. But if substitution occurs in one of the phenyl groups, the two forms regularly appear.[8]

The dioximes.—When the peculiar oxime grouping occurs several times in the molecule, the number of the isomers rises, amounting to three when the formula is symmetrical, as with benzildioxime, $(C_6H_5CNOH)_2$. Dioximidosuccinic acid,

$$(CO_2HCNOH)_2,$$

camphor-, anisyl-, nitrobenzil-, and ditolyl-dioxime, probably also the simplest glyoxime, $(HCNOH)_2$, and phenylglyoxime,[9] occur in two forms. Recent

[1] Goldschmidt, *Ber.* **26**, 2084. [2] Hantzsch, *ibid.* **24**, 5, 3479.

[3] Jovitschitsch, *ibid.* **28**, 2683.

[4] Hirsch, *Monatsh. f. Chem.* **16**, 831.

[5] *Ber.* **24**, 41. [6] Dollfus, *ibid.* **25**, 1932.

[7] Lossen, *Ann.* **175**, 271; **186**, 1; **252**, 170.

[8] With the oximes must be ranked the anil compounds, $XYC:NC_6H_5$, since v. Miller and Plöchl (*Ber.* **27**, 1296) have prepared, by the action of acetaldehyde on aniline, two isomeric ethylidene anilines $(CH_3.HC:NC_6H_5)_2$. See also *Ber.* **29**, 1733. L. Simon, however, attempted in vain to prepare these isomers (*Bull. Soc. Chim.* [3], **13**, 334).

[9] *Ber.* **24**, 25.

additions to this list are the dioximes of quinone and thymoquinone.

The facts concerning the oximes are, then, very simple: regular occurrence of two isomers[1] for compounds of the formula XYCNOH; disappearance of this isomerism when X and Y become identical; increase in the number of isomers when the above-mentioned group occurs more than once in the same molecule.

The observations concerning allied bodies may now be given. The groups to be considered where nitrogen occurs doubly-linked with carbon are these:

The hydrazones[2] and carbazides.[3]—Just as oximes are formed by the action of hydroxylamine on aldehydes or ketones, &c., *i.e.* on compounds containing the group CO, so the hydrazones are formed by a corresponding action of hydrazines on these compounds. And if the group CO is first replaced by CCl_2, there are again formed two isomers, provided the groups linked with carbon are different. The

[1] There are, however, many exceptions—cases in which only one isomer has been isolated. There is only one oxime of pyruvic acid, of thienylglyoxylic acid, of the ortho-substituted aromatic acids, of the mixed ketones containing an aliphatic and an aromatic radical (Claus and Häfelin, *J. prakt. Chem.* **54**, 391). But if we compare this single oxime with two stereomers of analogous constitution and of known configuration, we find that in its chemical and physical properties it resembles one of the two isomers, and totally differs from the other. These are, then, extreme cases of the instability of one isomer.

[2] Fehrlin and Krause, *Ber.* **23**, 1574, 3617; Hantzsch and Kraft, *ibid.* **24**, 3511; Hantzsch and Overton, *ibid.* **26**, 9, 18.

[3] Marckwald, *ibid.* **24**, 2880; Dixon, *J. Chem. Soc. Trans.* 1892, 1012.

substances at present known have been prepared from phenyl- and diphenyl-hydrazine, and correspond therefore to the formulæ

$$XYCNNHC_6H_5 \quad \text{and} \quad XYCNN(C_6H_5)_2.$$

The following derivatives of this kind have been obtained in two forms:

Phenylhydrazone of	$XYCNNHC_6H_5$
o-Nitrophenylglyoxylic acid	$C_6H_4NO_2(CO_2H)CNNHC_6H_5$
Anisylphenylketone	$C_6H_5(C_6H_4OCH_3)CNNHC_6H_5$
Carbazide	$C_6H_5NH(SH)CNNHC_6H_5$
p-Tolylcarbazide	$C_7H_7NH(SH)CNNHC_6H_5$
Phenyl-p-tolylcarbazide	$C_6H_5NH(SH)CNNHC_7H_7$
o-Tolyl-p-tolylcarbazide	$C_7H_7NH(SH)CNNHC_7H_7$
Di-p-tolylcarbazide	„ „
Benzylphenylcarbazide	$C_7H_7NH(SH)CNNHC_6H_5$
Diphenylhydrazones of	$XYCNN(C_6H_5)_2$
Anisylphenylketone	$C_6H_5(C_6H_4OCH_3)NN(C_6H_5)_2$
Tolylphenylketone	$C_6H_5(C_6H_4CH_3)NN(C_6H_5)_2$

The carbodi-imides.—By abstracting hydrogen sulphide from sulphocarbanilide, $SC(NHC_6H_5)_2$, Weith[1] obtained a carbodiphenylimide, $C(NC_6H_5)_2$, which according to Schall's[2] researches occurs in three modifications of equal molecular weight. According to Miller and Plöchl,[3] however, there are only two modifications, of which one has thrice the molecular weight of the other.

The diazotates.

$$C_6H_5.N_2.X.(X = Cl, OMe, SO_3Me, CN).$$

Besides the structural isomers,

$$(1)\ C_6H_5.\underset{\displaystyle N}{\overset{}{N}}.X \quad \text{and} \quad (2)\ C_6H_5N:NX,$$

[1] *Ber.* **7**, 1306.

[2] *Ibid.* **25**, 2880; **26**, 3064; *Zeitschr. physik. Chem.* **12**, 145.

[3] *Ber.* **28**, 1004.

there is evidence[1] to show that the compounds possessing the formula (2) exist in two forms, to which Hantzsch attributes the stereomeric formulæ

$$\begin{array}{ccc} C_6H_5\underset{\cdot\cdot}{N} & & C_6H_5\underset{\cdot\cdot}{N} \\ X\dot{N} & \text{and} & \dot{N}X \\ \text{syn-diazotate} & & \text{anti-diazotate} \end{array}$$

The question of the constitution of these substances is still in dispute.[2]

The isomerism of the bodies $H_2N_2O_2$ is also attributed by Hantzsch [3] to doubly linked nitrogen:

$$\begin{array}{cccc} & HO.N & & HO.N \\ (1) & HO.\ddot{N} & (2) & \ddot{N}.OH \\ & \text{syn-} & & \text{anti-hyponitrous acid} \end{array}$$

The isomer (1) is that which readily breaks up into N_2O and water.

C. TRIVALENT NITROGEN IN CLOSED RINGS

Just as in the case of carbon the double linkage and the fumar-maleïc isomerism were treated in connection with ring linkage and the isomerism of the hydrophthalic acids, so here some remarkable observations by Ladenburg and by Giustiniani should be mentioned.

The former, having shown that there are probably five isomeric piperidinemonocarboxylic acids,[4]

[1] Hantzsch, *Ber.* **28**, 1734; Hantzsch and Gerilowski, *ibid.* **28**, 2002; **29**, 743, 1059.

[2] Bamberger, *Ber.* **29**, 564, 1388; Blomstrand, *J. prakt. Chem.* **54**, 305.

[3] *Ann.* **292**, 340; Hantzsch and Kaufmann, *ibid.* **292**, 317.

[4] *Ber.* **25**, 2775.

proved[1] that conine ($\alpha_D = 13{\cdot}8°$), on heating the chlorhydrate with zinc dust, is transformed into an isomer ($\alpha_D = 8{\cdot}2°$),[2] and that this was also the case with the active α-methylpiperidine.[3]

As the rotation alone indicates, it may be supposed that this is not a case of isomerism caused by the asymmetric carbon atom.

Giustiniani[4] found that benzylmalimide occurs in two isomeric forms, of which one is distinguished from the other by having about double the rotation :

α-Imide	$\alpha_D = -24{\cdot}3°$ (2·28 %)	β-Imide	$\alpha_D = -48{\cdot}2°$ (2·255 %)
„	$\alpha_D = -21{\cdot}4°$ (0·244 %)	„	$\alpha_D = -43°$ (0·226 %)

This isomerism is maintained in the acetyl and benzoyl derivatives, but is lacking in the benzylmalamic acid, which is formed by treatment with potash.

If we compare the constitutional formulæ

$$\mathrm{HC}\left\langle\begin{matrix}\mathrm{CH_2{-}CH_2}\\ \\ \mathrm{CH_2{-}\mathit{C}HX}\end{matrix}\right\rangle\mathrm{NH} \qquad \begin{matrix}\mathrm{H_2C{-}CO}\\ | \\ \mathrm{HOH\mathit{C}{-}CO}\end{matrix}\Big\rangle\mathrm{NC_7H_7}$$

we find here a structure analogous to the above cases of double linkage, since here symmetry is lacking in the carbon radical attached to the nitrogen. Perfectly analogous cases of isomerism have been recently

[1] *Preuss. Akad.* 1892, 1057.

[2] Wolffenstein accounts this a mixture of inactive and dextro-conine (*Ber.* **27**, 2616 ; **29**, 195). But see Ladenburg, *ibid.* **29**, 2706.

[3] But see Marckwald, *Ber.* **29**, 43, 1293 ; and Ladenburg, *l.c.* p. 422.

[4] *Gazz. Chim.* 1893, 168.

discovered by Ladenburg[1] among the imides of tartaric acid.

D. CONFIGURATION IN THE CASE OF DOUBLY LINKED NITROGEN[2]

Combining the two ideas, that from carbon there proceed four directive forces as divergent as possible—that is, directed towards the tetrahedron corners—and also from nitrogen three forces in one plane directed towards the corners of a triangle, we arrive at the annexed fig. 18 in the same way as we deduced the figure for doubly linked carbon. The essence of this arrangement is that, as in ethylene derivatives, all the components, ANCXY, must be arranged in one plane. Under the influence of the attractive forces proceeding from X and Y, the radical at A appears to find its position of equilibrium either nearer to X or nearer to Y; but as one of these positions will be more favoured than the other, we can always distinguish a stable and a labile modification. Hantzsch and Werner represent this very suitably thus:[3]

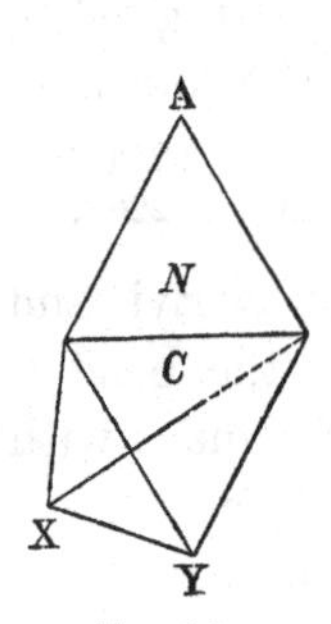

Fig. 18.

[1] *Ber.* **29**, 2710.

[2] Willgerodt, *J. prakt. Chem.* **37**, 449; Marsh, *J. Chem. Soc.* 1889, 654; Hantzsch and Werner, *Ber.* **23**, 11; Werner, *Räumliche Anordnung der Atome in stickstoffhaltigen Molekülen*, 1890; *Beiträge zur Theorie der Affinität und Valenz*, 1891; Vaubel, *Das Stickstoffatom*, 1891; V. Meyer and Auwers, *Ber.* **24**, 4229; **26**, 16.

[3] But compare Béhal, *Actualités Chimiques*, **1**, 76; Jovitschitsch, *l.c.* 167.

NA ‖ XCY and AN ‖ XCY

The three isomers of benzildioxime would then be represented thus:

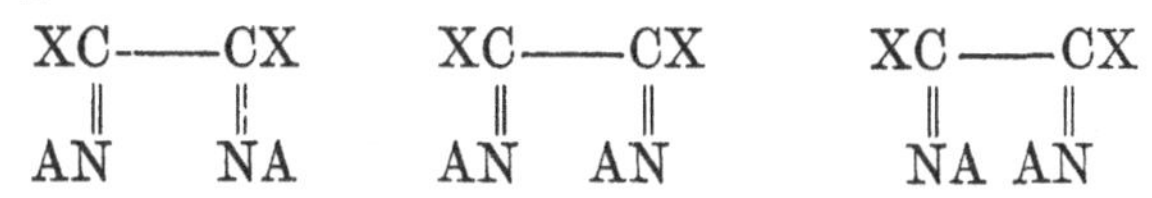

For the corresponding isomerism in ring compounds the relations would be expressed, according to Ladenburg, by the symbols used on p. 121, as follows:

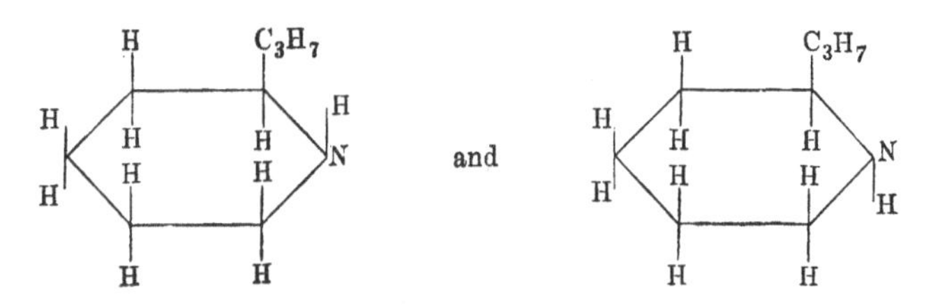

Nitrogen linked with nitrogen.—It must not be overlooked that Willgerodt[1] has observed in the case of the picrylhydrazines formed from dinitrochlorobenzene and phenylhydrazine,

$$C_6H_3(NO_2)HNNH(C_6H_5)$$

—as well as in the case of picryl-α- and -β-naphthylhydrazine,[2] which are constituted according to the formula R_1HNNR_2H—an isomerism which, on oxidising these compounds to the azo-derivative R_1NNR_2, disappears.

[1] *J. prakt. Chem.* **37**, 449.

[2] *Ibid.* **43**, 177.

The explanation given, which is based upon the difference between the symbols

$$\begin{array}{c} R_1NH \\ | \\ R_2NH \end{array} \quad \text{and} \quad \begin{array}{c} R_1NH \\ | \\ HNR_2 \end{array}$$

would indicate that free rotation is stopped by a single nitrogen-nitrogen linkage, as by a double carbon linkage.

The cause of this might be found in the supplementary valences, and doubly linked nitrogen would then be analogous to trebly linked carbon and cause no isomerism.

II. Compounds containing Pentavalent Nitrogen

Besides these researches on derivatives where the nitrogen is trivalent, there are some observations of Le Bel's on ammonium compounds. In the first place he succeeded[1] in obtaining two isomeric trimethylisobutylammonium chlorides, a result which calls to mind the isomerism which Ladenburg[2] stated to exist in the trimethylbenzyl derivative, but which Meyer[3] doubted. The isomerism discovered by Le Bel shows itself in the chloroplatinate, which at first forms in needles, but after recrystallisation from alcohol in octahedra. This second type is regained unaltered after treatment with silver oxide and reconversion to chloroplatinate; but if the experiment lasts some time the needles result on re-

[1] *Compt. Rend.* **110**, 144. [2] *Ber.* **10**, 43, 561, 1152, 1634.
[3] *Ibid.* 309, 964 978 (Corrp.), 1291.

formation of the platinate. One compound therefore is the more stable as chloroplatinate, the other as hydroxide. It is to be observed that if the substituted groups are smaller (trimethylpropyl, tripropylmethyl) the isomerism in question does not occur, probably in consequence of an intra-molecular transformation, which, in fact, is favoured by the mobility of the smaller groups. The same thing is observed among the oximes ; ethylaldoxime is easily transformed, and probably for the same reasons the simplest members of the ketoximes are lacking (*e.g.* phenylmethylketoxime). Schryver[1] made similar observations. While the corresponding ethyl and methyl derivatives showed no isomerism, it was found that on treating methylethylisoamylamine with ethyliodide a chloroplatinate results, which on warming is converted into the compound obtained direct from methyl iodide or amyl iodide and the appropriate amine. Here, then, we have isomerism in the case of

$$(H_3C)_3C_4H_9NCl \text{ and } (H_3C)(C_2H_5)_2C_5H_{11}NCl.$$

Finally, Le Bel[2] has made the most important observation, that isobutylpropylethylmethylammonium chloride may be 'doubled,' and yields active compounds, numbering probably four. The chlorides of ethylpropyldimethyl, ethyldipropylmethyl, ethyldipropylisobutyl, and ethylpropyldiisobutyl ammonium could not be 'doubled.'

The only conclusion at present to be drawn from

[1] *J. Chem. Soc. Proc.* 1891, 39. [2] *Compt. Rend.* **112**, 724.

what has just been said is that in ammonium chloride, in view of the activity among its derivatives, all the atoms do not lie in one plane; while in view of the isomerism of the trimethylisobutyl derivative, the four hydrogen atoms have not identical positions in the molecule. The inactivity observed when two groups are identical would indicate that the similar groups are symmetrically situated with regard to the plane passing through the two others, the nitrogen and the chlorine.

For graphic representation I will reproduce here that cube which I long ago proposed (p. 169). The nitrogen is supposed to be in the centre and the five connected groups in five of the corners (fig. 19). Of these 1, 2, and 3, which have equivalent positions, correspond to the alkyls attached to the three chief valences. When the nitrogen is trivalent they lie in one plane with it; here they are somewhat displaced through the influence of the chlorine situated in 4; in 5 lies the fourth alkyl.

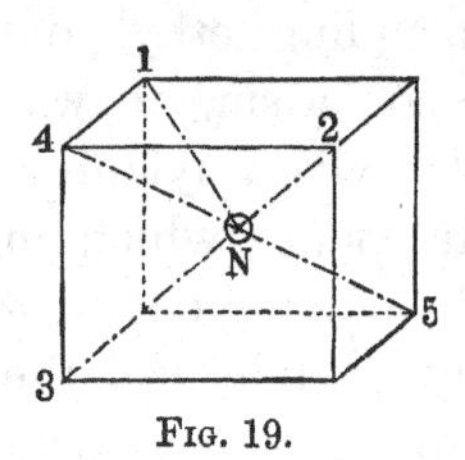

FIG. 19.

If one of the alkyls is different from the three others, which are identical, so that the type

$$(R_1)_3R_2NCl$$

results, as in $(H_3C)_3C_4H_9NCl$, there is the possibility of isomerism according as C_4H_9 is in 5 or in 1 to 3. And this isomerism has actually been observed. As yet there is no reason to expect optical activity.

The stability of one isomer has been found to be very slight.

If two different alkyls have entered the molecule, which would give the type $(R_1)_2R_2R_3NCl$, then besides asymmetric (*i.e.* active) configurations (R_1 in 5), there is always a symmetrical configuration possible (R_2 or R_3 in 5); and the slight stability above mentioned leads to the symmetrical type, which always corresponds to the favoured position of equilibrium. Accordingly 'doubling' has not succeeded here, *e.g.* in the case of

$$C_2H_5(CH_3)_2C_3H_7NCl.$$

If three different radicals have entered the molecule (type $R_1R_2R_3R_4NCl$), internal symmetry is impossible. The 'doubling' succeeded here in the case of $(C_4H_9)(C_3H_7)(C_2H_5)(CH_3)NCl$, of which already several isomers have been prepared. Of these, four types should exist, according as one or the other of the four different groups occupies 5; each of the four types would be divisible into two isomers of opposite activity.

APPENDIX

STEREOCHEMICAL ISOMERISM OF INORGANIC COMPOUNDS

NOTE BY ALFRED WERNER

Professor of Chemistry in the University of Zürich -

To facilitate the study of the stereochemical isomerism presented by certain classes of inorganic compounds, we must glance briefly at the constitution of these substances. They are molecular compounds whose constitution can hardly be represented with the aid of the idea of valence, unless we resort to several secondary hypotheses, each applicable to only a limited number of compounds.

The constitution of the molecular compounds may be established on the basis of a relation between those known as ammoniacal metallic compounds and the double salts, such as the double chlorides, fluorides, nitrites, &c. Indeed, the two extreme groups may be connected by a certain number of intermediate bodies of mixed character, thus forming a continuous series in which the molecular combinations of the first class gradually pass into the double salts.

Let us consider this remarkable transition in one of the most simple series. In the study of the compounds in question the fact that certain electro-negative radicals in the molecule behave in a peculiar, an abnormal manner, is of great importance. To emphasise this peculiarity, let us take a special case. We are acquainted with two ammoniacal compounds of cobalt, the one corresponding to the formula $Co(NH_3)_6Cl_3$, the other to the formula $Co(NH_3)_5Cl_3$. It is seen that the two bodies differ only by a molecule of ammonia, and yet their chemical properties are very different and characterised by the following reactions. On adding a solution of nitrate of silver to a solution of the first salt, it is found that the three atoms of chlorine are precipitated as silver chloride, a nitrate, $Co(NH_3)_6(NO_3)_3$, being formed. In the case of the second salt, the nitrate of silver precipitates only two atoms of chlorine, the third differs entirely in its chemical function; a chloronitrate, $Co(NH_3)_5{}^{Cl}_{(NO_3)_2}$, results.

This difference in reaction is observed also in the case of other reagents. Thus, when acted on by concentrated sulphuric acid, the first salt loses its three atoms of chlorine as hydrochloric acid, while the second in the same circumstances loses only two molecules of hydrochloric acid.

Thus the three chlorine atoms of the second salt have not the same chemical function; one of them behaves in a special way like the chlorine in certain organic compounds. Arrhenius' hypothesis of electrolytic dissociation accounts for this anomaly. The two atoms of chlorine which have the same properties as the chlorine in the ordinary chlorides (chloride of potassium, &c.) behave as *ions*, while the third does not.

As is well known, one of the factors of the electric conductivity of a saline solution is the number of ions which it contains; the properties of the two salts $Co(NH_3)_6Cl_3$ and $Co(NH_3)_5Cl_3$ indicated, then, that there would be a difference in the conductivity of the solutions of these compounds. Experiment confirms this prevision.

For a dilution of 1,000 litres the molecular conductivity of the first salt has been found equal to 432·6, and that of the second to 261·3.

There can then be no doubt that the first salt contains three atoms of chlorine identical in properties and acting as ions, while the second contains only two which can act in this way.

What chiefly interests us is to find the difference of constitution to which we should refer the various properties of the negative groups forming part of the molecules in question.

All the chemists who have worked at this subject, whatever their theories as to the constitution of the ammoniacal metallic compounds, consider this difference of constitution as the consequence of a different connection of the negative group with the metallic atom, which connection may be either direct or indirect.

When the connection is direct—that is, when the negative group is directly united with the metal—this group does not behave as an ion. When the connection is indirect—that is, when the negative group is united to the metal indirectly by means of ammoniacal molecules—this group behaves as an ion. The difference between the two kinds of connection is indicated by the following formula:

$$Co<^{Cl}_{NH_3Cl}$$

Although this way of looking at the constitution of these compounds does not harmonise very well with the ideas which we ordinarily hold concerning the state of salts in solution, it is so thoroughly confirmed by all the facts observed with regard to the class of ammonio-metallic compounds, that it is hardly possible to doubt it, and we shall adopt it in the following discussion.

One of the simplest series of bodies which we have to consider is that of the derivatives of bivalent platinum.

The platinum atom combines with four molecules of ammonia to form a compound, $Pt(NH_3)_4X_2$, the letter X representing a monovalent acid radical. The reactions of these salts, and their molecular conductivity, prove that the two acid groups act as ions; they represent the acid radicals of a salt of which the positive part is the radical $Pt(NH_3)_4$.

The second term of the series is a compound,

$$Pt(NH_3)_3X_2\,;$$

the old constitutional formula $Pt\begin{matrix}NH_3.NH_3.X\\NH_3X\end{matrix}$ is not at all in accord with the observed molecular conductivity, which indicates that only one of the chlorine atoms behaves as an ion. The formula should then be

$$Pt\begin{matrix}(NH_3)_3X\\X\end{matrix}$$

The third term of the series, $Pt(NH_3)_2X_2$, is found in two isomeric forms, the salts of platosammine and the salts of platosemidiammine. The formulæ attributed by Cleve and Jörgensen to these salts are the following:

$$Pt < \begin{matrix}NH_3.Cl\\NH_3.Cl\end{matrix} \quad \text{and} \quad Pt < \begin{matrix}NH_3.NH_3.Cl\\Cl\end{matrix}$$

Now, neither of these formulæ accounts for the chemical properties and the electric conductivities of these salts. Indeed, these substances no longer behave at all like salts of strong bases; but the chlorine, in such salts as

$$Pt\begin{matrix}(NH_3)_2\\Cl_2\end{matrix},$$

has properties analogous to those possessed by this element in chlorinated organic bodies; the electric conductivity approaches zero, and it is with great difficulty that any chemical reactions can be brought about. From all this it follows that their formulæ must be

$$Pt\begin{matrix}X_2\\(NH_3)_2\end{matrix},$$

the two negative radicals being in direct union with the platinum; in this case the ammonia molecules must be united similarly; the rational formula will be then:

$$Pt \begin{cases} NH_3 \\ NH_3 \\ X \\ X \end{cases}$$

The negative groups being attached by means of valences, as it is usually called, the ammonia molecules by means of secondary forces, I shall indicate this difference by saying that the molecules of ammonia are *co-ordinated*, that is to say, that they must be directly connected with the metallic atom, the platinum, although this linkage is not due to what are ordinarily called valences.

The next term of the series is a compound,

$$PtNH_3Cl_2ClR,$$

R representing a monovalent positive group; it is a double salt in the ordinary sense of the word. Supposing R to represent an atom of potassium, the formula would ordinarily be written thus: $Pt^{NH_3}_{Cl_2} + KCl$. But the substance does not behave at all as this formula would indicate; on the contrary, its molecular conductivity proves that it contains a complex radical, $Pt^{NH_3}_{Cl_3}$, which acts as a negative ion, the potassium being the electro-positive ion. The compound in question is a salt of a peculiar kind, of which the acid radical is the group $Pt^{NH_3}_{Cl_3}$ and the basic radical the potassium.

The final term of the series is the compound

$$PtCl_2 + 2KCl,$$

the addition product formed by platinous chloride and potassium chloride. This salt, again, has not the properties which the above formula assigns to it, the molecular conductivity proving beyond doubt that we have to do with a salt of which the acid radical is formed by the group $PtCl_4$, the basic radicals being the two potassium atoms; the negative ion is $PtCl_4$, the positive ions are K_2, and the rational formula is $(PtCl_4)K_2$.

To resume. These substances form the following series:

$$[Pt(NH_3)_4]Cl_2; \quad \left[Pt{(NH_3)_3 \atop Cl}\right]Cl; \quad \left[Pt{(NH_3)_2 \atop Cl_2}\right];$$

$$\left[Pt{NH_3 \atop Cl_3}\right]K; \quad [PtCl_4]K_2.$$

In this series we observe that all the compounds contain a special radical (PtA_4), of which the character varies with the nature of the groups A. In the first terms this radical has a basic function; in the middle of the series it is neutral, and in the last terms it has an acid function. The constitution of this radical PtA_4 is of much interest. Everything tends to show, as we have pointed out, that the four groups A are directly connected with the platinum atom. If these four radicals are in the same plane with the platinum atom the constitution of these complex radicals will be represented by

$${A \atop A} > Pt < {A \atop A}$$

Admitting this arrangement of groups in one plane, we get, when two of the radicals A are different from the other two, a case of geometrical isomerism expressed by the formulæ:

$${A \atop A_1} > Pt < {A_1 \atop A} \qquad \text{and} \qquad {A \atop A} > Pt < {A_1 \atop A_1}$$

We must refer to these theoretical formulæ certain cases of isomerism observed among the compounds of platinum.

One of the most characteristic examples is the isomerism of the salts of platosemidiammine, $Pt\genfrac{}{}{0pt}{}{X_2}{(NH_3)_2}$, with the salts of platosammine, $Pt\genfrac{}{}{0pt}{}{X_2}{(NH_3)_2}$.

The two series of compounds correspond to the same formula, $Pt\genfrac{}{}{0pt}{}{(NH_3)_2}{X_2}$, and we can explain their isomerism only by the stereochemical formulæ,

$$\genfrac{}{}{0pt}{}{X}{X}>Pt<\genfrac{}{}{0pt}{}{NH_3}{NH_3} \quad \text{and} \quad \genfrac{}{}{0pt}{}{X}{NH_3}>Pt<\genfrac{}{}{0pt}{}{NH_3}{X}$$

We can even allot the proper formula to each of the two series.

Let us assume that, say, the first formula represents the compounds of platosemidiammine, the second formula the compounds of platosammine,

$$\underset{\text{Chloride of platosemidiammine}}{\genfrac{}{}{0pt}{}{Cl}{Cl}>Pt<\genfrac{}{}{0pt}{}{NH_3}{NH_3}} \qquad \underset{\text{Chloride of platosammine}}{\genfrac{}{}{0pt}{}{Cl}{NH_3}>Pt<\genfrac{}{}{0pt}{}{NH_3}{Cl}}$$

then the analogous compounds formed by platinous chloride with pyridine will have the formulæ :

$$\genfrac{}{}{0pt}{}{Cl}{Cl}>Pt<\genfrac{}{}{0pt}{}{Py}{Py} \qquad \genfrac{}{}{0pt}{}{Cl}{Py}>Pt<\genfrac{}{}{0pt}{}{Py}{Cl}$$

By treating the chloride of platosemidiammine with pyridine, and the chloride of platosemidipyridine with ammonia, we obtain the same compound, $\left[Pt\genfrac{}{}{0pt}{}{(NH_3)_2}{Py_2}\right]Cl_2$, which we shall call a and which is formed thus :

$$\left.\begin{array}{l}\genfrac{}{}{0pt}{}{Cl}{Cl}>Pt<\genfrac{}{}{0pt}{}{NH_3}{NH_3}+Py_2=\left[\genfrac{}{}{0pt}{}{Py}{Py}>Pt<\genfrac{}{}{0pt}{}{NH_3}{NH_3}\right]Cl_2 \\ \genfrac{}{}{0pt}{}{Cl}{Cl}>Pt<\genfrac{}{}{0pt}{}{Py}{Py}+(NH_3)_2=\left[\genfrac{}{}{0pt}{}{NH_3}{NH_3}>Pt<\genfrac{}{}{0pt}{}{Py}{Py}\right]Cl_2\end{array}\right\} a$$

Similarly, by treating the chloride of platosammine with pyridine, and the chloride of platopyridine with ammonia, we obtain a compound, $\left[Pt{(NH_3)_2 \atop Py_2}\right]Cl_2$, differing from α, and which we shall call β; this is stereomeric with α.

It is formed thus:

$$\left.\begin{array}{l} {Cl \atop NH_3}>Pt<{NH_3 \atop Cl} + Py_2 = \left[{Py \atop NH_3}>Pt<{NH_3 \atop Py}\right]Cl_2 \\ {Cl \atop Py}>Pt<{Py \atop Cl} + (NH_3)_2 = \left[{NH_3 \atop Py}>Pt<{Py \atop NH_3}\right]Cl_2 \end{array}\right\}\beta$$

On warming the compounds α and β they lose ammonia and pyridine and change into compounds of the platosammine series, that is, into bodies corresponding to the general formula:

$${A \atop X}>Pt<{X \atop A}$$

On considering the formulæ of α and β, it will readily be seen that the substance α will undergo such a transformation, on losing a molecule of ammonia and a molecule of pyridine according to the equation:

$$\left[{NH_3 \atop NH_3}>Pt<{Py \atop Py}\right]Cl_2 = {NH_3 \atop Py} + {NH_3 \atop Cl}>Pt<{Cl \atop Py}$$

The compound α should yield, then, in this reaction a substance, $Pt.{Py \atop {NH_3 \atop Cl_2}}$

The compound β, on the other hand, could undergo the transformation into salts of the platosammine series in two different ways, either by losing two molecules of ammonia, or by losing two molecules of pyridine, as shown by the following equations:

$$\left[{Py \atop NH_3}>Pt<{NH_3 \atop Py}\right]Cl_2 = Py_2 + {Cl \atop NH_3}>Pt<{NH_3 \atop Cl}$$

$$\left[{Py \atop NH_3}>Pt<{NH_3 \atop Py}\right]Cl_2 = (NH_3)_2 + {Py \atop Cl}>Pt<{Cl \atop Py}$$

On warming β, then, we should obtain a mixture of two substances, $\frac{NH_3}{C} > Pt < \frac{Cl}{NH_3}$ and $\frac{Py}{Cl} > Pt < \frac{Cl}{Py}$

These reactions, which may be deduced from the stereochemical formulæ of the compounds of platosammine and of platosemidiammine, are in fact those which occur on warming the substances α and β, of which the formulæ are thus settled.

It can be shown that if we give to the platosammine salts the formula $\frac{A}{A} > Pt < \frac{X}{X}$, and to the platosemidiammine salts the formula $\frac{X}{A} > Pt < \frac{A}{X}$, we shall arrive at conclusions which are no longer in accord with the facts.

For, on adding to platosammine chloride,

$$\frac{Cl}{Cl} > Pt < \frac{NH_3}{NH_3},$$

two molecules of pyridine, and to platopyridine chloride, $\frac{Cl}{Cl} > Pt < \frac{Py}{Py}$, two molecules of ammonia, we should get a compound $\left[\frac{NH_3}{NH_3} > Pt < \frac{Py}{Py}\right]Cl_2$; but this compound could change into salts of the platosammine series in three different ways: first, by losing two molecules of ammonia; second, by losing two molecules of pyridine; third, by losing one molecule of ammonia and one molecule of pyridine; we should obtain, then, a mixture of the three substances, $Pt\frac{Py_2}{X_2}$, $Pt\frac{(NH_3)_2}{X_2}$, $Pt.\overset{Py}{\underset{X_2}{NH_3}}$. Now, this has never been observed in a single case, even when the amines of the compound $\overset{a_2}{\underset{X_2}{Pt}}b_2$ are quite analogous in character, *e.g.* ethylamine and propylamine.

The two isomeric series must, then, correspond with the following formulæ:

$$\begin{matrix} X \\ X \end{matrix} > Pt < \begin{matrix} A \\ A \end{matrix} \qquad\qquad \begin{matrix} X \\ A \end{matrix} > Pt < \begin{matrix} A \\ X \end{matrix}$$

Salts of platosemidiammine **Salts of platosammine**

The number of stereomeric compounds of dyad platinum is already considerable. A special interest attaches to the compounds of sulphurous acid, having the formulæ:

$$\begin{matrix} Cl \\ HO_3S \end{matrix} > Pt < \begin{matrix} NH_3 \\ NH_3 \end{matrix} \quad \text{and} \quad \begin{matrix} Cl \\ NH_3 \end{matrix} > Pt < \begin{matrix} NH_3 \\ SO_3H \end{matrix}$$

and to the compounds

$$\begin{matrix} HO_3S \\ HO_3S \end{matrix} > Pt < \begin{matrix} NH_3 \\ NH_3 \end{matrix} \quad \text{and} \quad \begin{matrix} HO_3S \\ NH_3 \end{matrix} > Pt < \begin{matrix} NH_3 \\ SO_3H \end{matrix}$$

The substances thus far considered contain a radical MA_4. There exists also a large number of inorganic compounds whose molecules are characterised by the presence of a radical MA_6, and which may be arranged in series having characters analogous to those which we have developed in detail for dyad platinum.

To give an idea of these series, here are the formulæ of the compounds of tetrad platinum and of dyad cobalt:

$$[Pt(NH_3)_6]Cl_4, \quad \left[Pt\begin{matrix} Cl \\ (NH_3)_5 \end{matrix}\right]Cl_3, \quad \left[Pt\begin{matrix} Cl_2 \\ (NH_3)_4 \end{matrix}\right]Cl_2,$$

$$\left[Pt\begin{matrix} Cl_3 \\ (NH_3)_3 \end{matrix}\right]Cl, \quad \left[Pt\begin{matrix} Cl_4 \\ (NH_3)_2 \end{matrix}\right], \quad \left[Pt\begin{matrix} Cl_5 \\ NH_3 \end{matrix}\right]K,$$

$$[PtCl_6]K_2 \quad Co(NH_3)_6Cl_3, \quad \left[Co\begin{matrix} NO_2 \\ (NH_3)_5 \end{matrix}\right]Cl_2,$$

$$\left[Co\begin{matrix} (NO_2)_2 \\ (NH_3)_4 \end{matrix}\right]Cl, \quad \left[Co\begin{matrix} (NO_2)_3 \\ (NH_4)_3 \end{matrix}\right], \quad \left[Co\begin{matrix} (NO_2)_4 \\ (NH_3)_2 \end{matrix}\right]K,$$

$$\left[Co\begin{matrix} (NO_2)_5 \\ NH_3 \end{matrix}\right]K_2, \quad \left[Co(NO_2)_6\right]K_3.$$

Just as in the radicals MA_4 the four groups A are

directly connected with the atom of the metal, so in the compounds containing the complex radical MA_6 the six groups are in direct union with the metal; the proof is afforded by the amount of the molecular conductivity.

We have now to get an idea of the configuration of these groups MA_6; the most simple hypothesis that can be formulated is an octahedral arrangement; the metallic atom occupying the centre of the octahedron, the six groups A will have their places at the corners.

It is evident that this arrangement should give rise to certain cases of stereomerism, of which we shall consider at present only one, which experiment confirms.

Let us consider a radical MA_6 of which four groups are alike and the two others different: we have then a group $\left[M{A_4 \atop A'_2}\right]$. In this case the two radicals A′ may occupy different positions; they may occupy two corners of the octahedron joined by an axis, or two corners joined by an edge, as the following figures show:

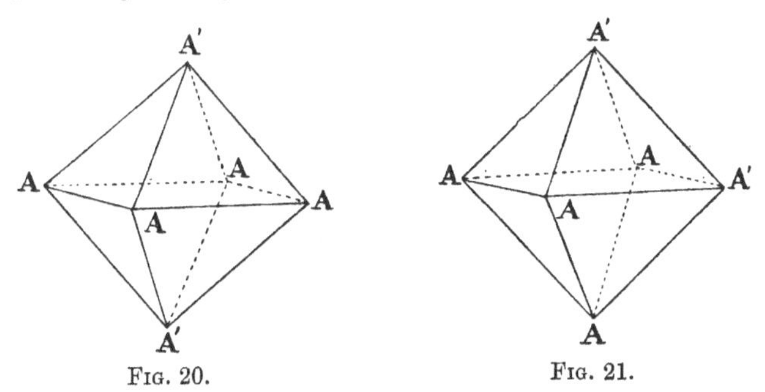

Fig. 20. Fig. 21.

that is to say that the compounds containing a radical $\left[M{A_4 \atop A'_2}\right]$, should present two isomeric forms.

The radical $\left[M{A_4 \atop A'_2}\right]$ is found in certain ammoniacal derivatives of cobalt, salts of praseocobaltammine, answering to the general formula $Co{}^{X_2}_{(NH_3)_4}X$; these salts should, then, if our theory is correct, present a special isomerism. And, as a matter of fact, this is what we find. We know by the beautiful researches of Jörgensen that there exist two series of salts of the formula $\left[Co{}^{X_2}_{A_4}\right]X$. The two series scarcely differ, from a chemical point of view; of the three acid radicals, only one acts as an ion. But the two series are distinguished by a characteristic property; the salts of the praseocobaltammine series are green, while the salts of the isomeric series, the salts of the violeocobaltammines, are violet, as their name indicates (see figs. 22 and 23).

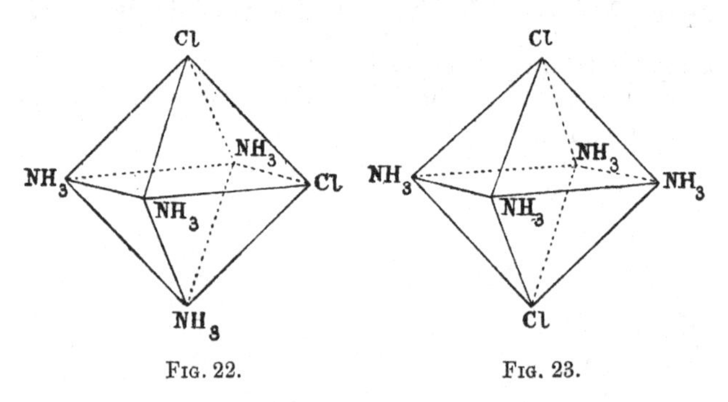

FIG. 22. FIG. 23.

This interesting case of isomerism is a first proof in favour of the stereomerism of the radicals $\left[M{}^{X_2}_{A_4}\right]$. In another series, also, cobalt presents this special isomerism. For a long time there has been known a group of salts of cobaltammine, called salts of crocéocobaltammine, and answering to the formula $\left[Co{}^{(NO_2)_2}_{(NH_3)_4}\right]X$; these also, then, contain a radical $\left[M^{X_2}_{A_4}\right]$. Quite recently Jörgensen has discovered a new series of compounds having the same formula, $\left[Co^{(NO_2)_2}_{(NH_3)_4}\right]X$, and differing from the first only in physical properties. He calls them salts of flaveocobaltammine, and it is impossible to doubt that this isomerism of the two series arises from the presence of two isomeric radicals, $\left[Co^{(NO_2)_2}_{(NH_3)_4}\right]$. To represent the positions occupied by the two NO_2 groups, one may imagine the following formulæ :

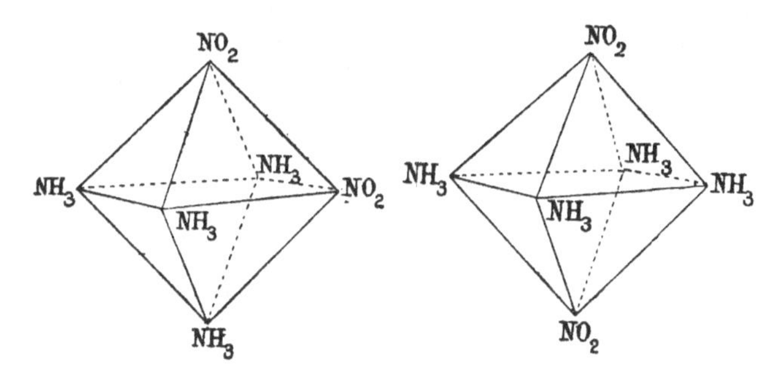

Among the ammoniacal derivatives of tetrad platinum we find a case of isomerism perfectly analogous to those observed among the cobalt compounds.

We know, in fact, two series of bodies answering to the general formula $Pt^{(NH_3)_2}_{X_4}$; they are the salts of platinosemidiammine and the salts of platinammine; here again, then, we encounter the radical $\left[M^{A_4}_{A'_2}\right]$.

Here, too, the isomerism is doubtless due to the same cause as with the cobalt compounds, and would be represented by the following formulæ :

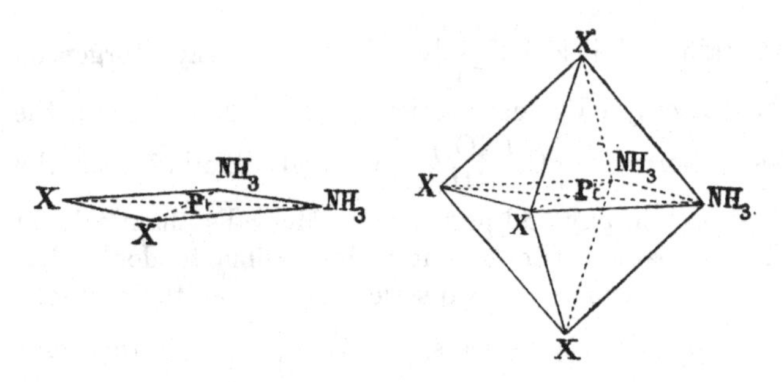

We can even determine, with a certain degree of probability, the space formula corresponding to each of the two series.

The compounds of the platinosemidiammine series and of the platinammine series are formed by the addition of two negative groups to the salts of platosemidiammine and of platosammine, the dyad platinum transforming itself into tetrad platinum :

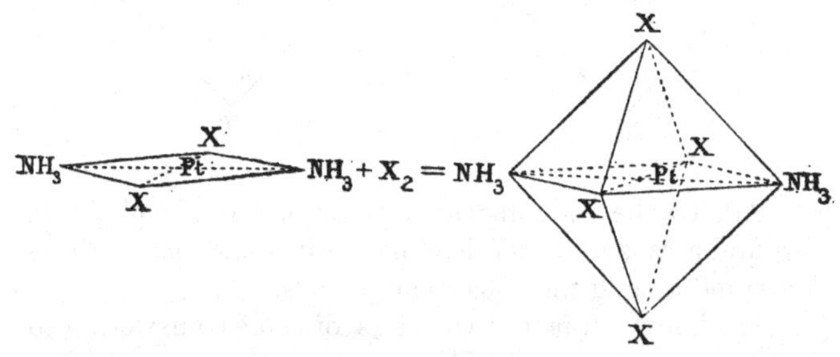

For the compounds of bivalent platinum we have arrived at plane formulæ; for those of tetravalent platinum we have given octahedral formulæ. The most simple hypothesis is, then, that the negative groups add themselves to the salts of divalent platinum, so as to occupy two corners united by the diagonal of an octahedron, which is formed by the four radicals joined to the platinum, and by the two added radicals which complete the molecule.

This transformation is explained by the above formulæ, which also give us the stereochemical formulæ of the two isomeric series.

In the short sketch here given of the stereochemical isomerism of certain classes of inorganic compounds we have been able to consider only the principal points of the new theory; we believe, however, that we have proved, by well-established facts, that it is possible to explain these cases of isomerism only by stereochemical conceptions.

Bibliography

A. Werner, 'Contribution to the Constitution of Inorganic Compounds,' *Zeitschr. f. anorg. Chem.* **3**, 267.

A. Werner and A. Miolati, *Zeitschr. f. physik. Chem.* **12**, 35; **13**, 506.

INDEX

Errata.

Page 52, line 11 from top, for *ammonia* read *ammonium*.
Page 87, the formula of lyxose should be as on page 83.
Page 95, line 12 from bottom, for *styrol* read *styrolene*.
Page 101, line 4 from top, for *nitrostyrol* read *nitrostyrolene*.
Page 166, line 3 from bottom, for *limonenenitroso chloride* read *limonene nitrosochloride*.

PRINTED BY
SPOTTISWOODE AND CO., NEW-STREET SQUARE
LONDON

Printed in the United States
By Bookmasters